Lecture Notes in Statistics

Edited by J. Berger, S. Fienberg,
J. Gani, and K. Krickeberg

44

D.L. McLeish
Christopher G. Small

The Theory and Applications of Statistical Inference Functions

Springer-Verlag

New York Berlin Heidelberg London Paris Tokyo

D.L. McLeish
Christopher G. Small
Department of Statistics and Actuarial Science
University of Waterloo
Ontario, Canada N2L 3G1

Mathematics Subject Classification (1980): 62C12

Library of Congress Cataloging-in-Publication Data
McLeish, D. L.
  The theory and applications of statistical inference
functions.
  (Lecture notes in statistics ; 44)
  Bibliography: p.
  Includes index.
  1. Mathematical statistics. I. Small, Christopher G.
II. Title. III. Series: Lecture notes in statistics
(Springer-Verlag) ; v. 44.
QA276.M39 1988   519.5   88-6431

Camera-ready text prepared by the authors.

9 8 7 6 5 4 3 2 1

ISBN-13: 978-0-387-96720-2   e-ISBN-13: 978-1-4612-3872-0
DOI: 10.1007/978-1-4612-3872-0

Preface

This monograph arose out of a desire to develop an approach to statistical inference that would be both comprehensive in its treatment of statistical principles and sufficiently powerful to be applicable to a variety of important practical problems. In the latter category, the problems of inference for stochastic processes (which arise commonly in engineering and biological applications) come to mind. Classes of estimating functions seem to be promising in this respect. The monograph examines some of the consequences of extending standard concepts of ancillarity, sufficiency and completeness into this setting.

The reader should note that the development is mathematically "mature" in its use of Hilbert space methods but not, we believe, mathematically difficult. This is in keeping with our desire to construct a theory that is rich in statistical *tools* for inference without the difficulties found in modern developments, such as likelihood analysis of stochastic processes or higher order methods, to name but two. The fundamental notions of orthogonality and projection are accessible to a good undergraduate or beginning graduate student. We hope that the monograph will serve the purpose of enriching the methods available to statisticians of various interests.

We are indebted to Prof. Mary Thompson for her useful comments on an early version of this manuscript. Each of us would also like to thank the other author for typing the manuscript. Naturally, any errors are the responsibility of the other author.

Contents

CHAPTER 1
INTRODUCTION

The theory of estimating functions has become of interest in a wide variety of statistical applications, partly because it has a number of virtues in common with methods such as maximum likelihood estimation while possessing sufficient flexibility to tackle problems where maximum likelihood fails, such as the Neyman-Scott paradox. In this monograph we present a self-contained development with a number of applications to estimation, censoring, robustness and inferential separation of parameters.

A basic concept of the theory of estimating equations (or inference functions as we shall call them here) involves a change of emphasis. Attention is turned from the task of estimating a parameter directly to the problem of estimating an unknown function which possesses the true value of the parameter as a root. That there are virtues in this perspective is by no means obvious, as the approach to inference would seem to be quite oblique. However, let us examine the use of maximum likelihood estimation as enshrined in statistical practice. Much of the theory of inference functions can be found in the logic underlying asymptotic theory of maximum likelihood estimation. Usually, a maximum likelihood estimate $\hat{\theta}$ is adopted as an estimator for philosophical or practical reasons and is found to possess certain desirable properties such as consistency or efficiency in the presence of regularity conditions imposed on the model. However, there are a number of obstacles in these arguments.

I. In practice we must distinguish between the root of the score function $S(\theta) = \frac{\partial}{\partial \theta}\log f(\theta;x)$ and the m.l.e. $\hat{\theta}$. For example, the likelihood may be maximized on the boundary of the parameter space so that $\hat{\theta}$ is distinct from any root of $S(\theta)$. In many cases, proofs of consistency for maximum likelihood estimates are more properly proofs of the existence of a consistent root of the score function. The consistency of the global maximum of the likelihood can be achieved through the conditions of Wald(1949). However, such a result is obtained at the expense of the generality and simplicity of those consistency proofs depending solely upon the law of large numbers and smoothness of the likelihood function. As Lehmann(1983; p. 420) illustrates in an example due to LeCam, the global maximum of the likelihood can be inconsistent while at the same time a local maximum is consistent. If we examine the argument for consistency we see that it contains two steps. It is first shown that $S(\theta)$ converges to $E_{\theta_0} S(\theta)$ for every θ and then this convergence is "transferred" across to the estimator $\hat{\theta}$ using the fact that $S(\hat{\theta}) = 0$ and $E_{\theta_0} S(\theta_0) = 0$. It is the second step which is the more problematical of the two because a regularity condition such as monotonicity of $S(\theta)$ is needed to establish the result. The first step holds from the law of large numbers under fairly general conditions in an i.i.d. model.

II. Asymptotic normality of maximum likelihood estimates is usually proved by first proving normality of the score function $S(\theta)$ and then using the local linearity of the function in a neighborhood of the true value of θ. It can be seen that there is a logical order to these two steps and that the second step (local linearity) is often the weakest part of the proof for small samples. In the product model of i.i.d. observations, the score function is just the sum of the score functions for individual observations. So the central limit theorem applies directly. However, maximum likelihood estimates are not in general the average of the estimates based upon individual observations. The result is that the score function is often "more normal" than the estimator defined by its root.

III. The distinction between the asymptotic normality of $S(\theta)$ and the asymptotic normality of $\hat{\theta}$ is closely linked to the concept of parametrization invariance. Let $\theta_1 = \eta(\theta)$ be a smooth reparametrization with non-vanishing derivative. In the new parameter θ_1 we can write

$$S_1(\theta_1) = S(\eta^{-1}(\theta_1))\frac{d\theta}{d\theta_1} \tag{1.1}$$

Thus S_1 will be a scale multiple of S and it will follow that under any scale-invariant measure of non-normality the distributions of S and S_1 will be equally close to normal. We can summarize this argument by stating that the asymptotic normality of $S(\theta)$ is parametrization invariant. The same is not true of the maximum likelihood estimate $\hat{\theta}$. Although maximum likelihood estimation is parametrization invariant in that $\hat{\theta}_1 = \eta(\hat{\theta})$, the asymptotic normality of $\hat{\theta}$ is not invariant. For example, $\hat{\theta}$ can be exactly normally distributed while $\hat{\theta}_1$ is far from normal.

Arguments I, II and III above all lead to the same conclusion, namely that the asymptotic properties of $\hat{\theta}$ are obtained through the asymptotics of $S(\theta)$ which are transferred to the parameter space via local linearity. This in itself is enough to recommend the properties of $S(\theta)$ as being of independent interest. It turns out that a number of common procedures can be decomposed in a similar way. Marginalization provides an example. Suppose T is a function of the complete data set X. In incomplete data problems, T will be the observed data and X the unobservable complete data set. One approach to estimation of θ with incomplete data is to marginalize to T and to maximize the likelihood based upon the marginal distribution of T. Let X have p.d.f. $f(\theta;x)$ and T have p.d.f. $g(t;\theta)$ We write

$$S(\theta) = \frac{\partial}{\partial\theta}\log f(\theta;x)$$

and

$$S_T(\theta) = \frac{\partial}{\partial\theta}\log g(t;\theta)$$

The complete data estimate will be $\hat{\theta}$, where $S(\hat{\theta}) = 0$ and the incomplete data estimate will be called $\hat{\theta}_T$ where $S_T(\hat{\theta}_T) = 0$. Now an important result of Fisher(1925) states that

$$S_T(\theta) = E_\theta[\, S(\theta) \mid T].\qquad(1.2)$$

Thus $S_T(\theta)$ is seen to be the projection of $S(\theta)$ into the space of $\sigma(T)$-measurable functions. So the derivation of the maximum marginal likelihood estimate $\hat{\theta}_T$ can be decomposed into the projection of $S(\theta)$ into the space of $\sigma(T)$-measurable functions and the extraction of a root of the resulting image function. Now our concern is whether the score function $S_T(\theta)$ is in some sense a more fundamental or natural object than the estimator $\hat{\theta}_T$ that it defines. Let us suppose that on the basis of complete data, a maximum likelihood estimator $\hat{\theta}$ has been found. Suppose that $\hat{\theta}$ is a function of T. One might suppose that $\hat{\theta}_T = \hat{\theta}(T)$. However, this is *not* the case in general as T need not be sufficient. The failure of this identity contrasts with identity (1.2) and shows that score functions have properties not shared by the estimators that they define. This shows that there is a simple relationship between the score functions of the complete and incomplete data problems; there is no such simple relationship between the corresponding roots of these equations.

In view of the decompositions mentioned above it is reasonable to consider whether efficiency criteria for inference will also be effective if applied to inference functions rather than estimates in view of regularity requirements. For example we can impose efficiency criteria $eff(\psi;\theta)$ on a space of inference functions ψ. This approach, suggested by Godambe(1960), selects ψ so as to maximize the efficiency $eff(\psi;\theta)$. If $\psi^*(\theta)$ maximizes $eff(\psi;\theta)$ for every parameter value θ, then an estimate $\hat{\theta}^*$ could be chosen so that $\psi^*(\hat{\theta}^*) = 0$. The hope would be that such a theory would circumvent some of the regularity requirements of maximum likelihood estimation and provide a satisfactory theory of inference under more general circumstances. Godambe(1960) has developed such a theory, but as we shall see below, this measure of efficiency has its own difficulties with regularity.

Consider

$$E_\theta S(\theta;X) = E_\theta \frac{\partial}{\partial \theta} \log f(\theta;X) \qquad (1.3)$$

$$= \int \frac{\partial}{\partial \theta} f(\theta;x) dx$$

$$= \frac{\partial}{\partial \theta} \int f(\theta;x) dx$$

$$= \frac{\partial}{\partial \theta}(1) = 0.$$

The argument holds provided there is sufficient regularity to ensure that derivatives and integrals can be exchanged above. If we define

$$H(\theta) = E_{\theta_0} S(\theta;X)$$

where θ_0 is the true value of the parameter, then $H(\theta_0) = 0$. As S is an estimator of its expectation, H, we can regard a root $\hat{\theta}$ of S as an estimator of the root θ_0 of H. A similar argument will hold with S replaced by $\psi(\theta) = X - E_\theta(X)$, when $X \epsilon R$.

An inference function ψ is said to be unbiased if $E_\theta \psi(\theta;X) = 0$ for every θ in the parameter space. It may be that θ does not have an unbiased estimator. However, as we have seen above, unbiased inference functions exist under fairly general circumstances. Godambe(1960) has suggested the efficiency criterion

$$eff(\psi;\theta) = \{E_\theta \frac{\partial}{\partial \theta} \psi(\theta;X)\}^2 / E_\theta \psi^2(\theta;X) \tag{1.4}$$

and has shown that within a wide class of unbiased inference functions, eff is maximized by inference functions of the form

$$\psi^*(\theta;X) = k(\theta)S(\theta;X) \tag{1.5}$$

where $k(\theta)$ is arbitrary. However, the following example illustrates that this efficiency criterion has its own serious problems with non-regular examples.

1.1 Example. Suppose X has distribution which is uniform on $[0,\theta]$. We wish to estimate the parameter θ. Note that the score function $S(\theta;x) = \frac{d}{d\theta} \log f(\theta;x)$ is defined for $x < \theta$ and equals $\frac{-1}{\theta}$ there. Thus, there is no version of this derivative which is unbiased; the usual regularity conditions insuring that $E_\theta S(\theta;X) = 0$ fail, in this case. However, there is a natural extension of the score function which is well defined in this problem. Define a *score functional* S_θ by

$$S_{\theta}\psi = \frac{d}{d\xi}\{E_{\xi}\psi(\theta;X)\}|_{\xi=\theta} \qquad (1.6)$$

as a functional on unbiased estimating functions ψ for which the right hand side is well defined. Under the usual regularity conditions allowing differentiation under the integral,

$$S_{\theta}\psi = E_{\theta}\left[-\frac{d}{d\theta}\psi(\theta;X)\right] \qquad (1.7)$$

$$= E_{\theta}\{S(\theta;X)\psi(\theta;X)\}$$

In other words, under regularity conditions, the score functional can be represented as an inner product in $L_2(\theta)$. However, it is well known that such a representation is possible for a linear functional on a Hilbert space only if that functional is bounded, in which case the existence of the score function is given by the Riesz representation theorem. In many interesting problems such as the one we are treating here, this is not the case. Here, for left continuous ψ,

$$S_{\theta}\psi = \frac{1}{\theta}\psi(\theta;\theta)$$

It is easy to see that this score functional is not bounded. In particular, the unbiased estimating functions defined for all $1 < n < \infty$ and all $1 < r < \infty$ by

$$\psi_{n,r}(\theta;x) = \begin{cases} -C(1-rx/\theta)^r & x \le \dfrac{\theta}{r} \\[2mm] (\dfrac{x}{\theta}-\dfrac{1}{r})^n & x \ge \dfrac{\theta}{r} \end{cases}$$

are such that $\psi_n(\theta;\theta) = (1-\frac{1}{r})^n$, the function is continuously differentiable. In order that the function be unbiased, we must choose

$$C = \frac{r(r+1)}{n+1}\left(1-\frac{1}{r}\right)^{n+1}$$

and it is easy to see that if $r \to \infty$ and $r/n \to 0$,

$$E_\theta \psi_{n,r}^2(\theta;X) \approx \left(1-\frac{1}{r}\right)^{2n}\left[\frac{r^2}{2n^2}+\frac{1}{2n}\right]$$

Consequently, $eff(\psi_{n,r};\theta) = \dfrac{(S_\theta \psi_{n,r})^2}{E(\psi_{n,r}^2)} \to \infty$ as $n,r \to \infty$. The estimators corresponding to these estimating functions are obtained from the equations $\psi_{n,r}(\hat\theta;X) = 0$ and the solution satisfies $\hat\theta \to \infty$. However, a more reasonable estimator such as the maximum likelihood estimator can be made to correspond to the "least efficient" end of the scale, i.e. the root as $n \to 1$. This is a consequence of the failure of the regularity conditions which lead to unbiasedness of the score function for maximum likelihood estimation. The example is presented here simply to indicate that the order induced by norms such as Godambe's efficiency measure do not escape from the problems experienced by maximum likelihood estimators when the usual regularity conditions guaranteeing unbiasedness of the score function fail. In these sorts of models, sufficiency reduction is reasonable, and it is our purpose to explore a richer theory on the estimating functions parallel to standard concepts such as sufficiency.

In Chapter 2 we shall introduce the concepts of E-ancillarity and E-sufficiency which permit inferential reduction analogous to the usual sufficiency and ancillarity reductions but within the class of unbiased inference functions. The theory will be sufficiently general to encompass the non-regular example above. In Chapter 3, where we discuss criteria for the selection of an inference function we will see that regularity will still have an important role. However, we will turn the usual argument on its head:

instead of investigating regular models for which a specific inference function has satisfactory properties, we shall investigate the selection of a regular function $\psi(\theta)$ within an E-sufficient class. In other words, regularity will be demanded of the function rather than the model. This is in fact a more practical approach in as much as the choice of function is under the control of the statistician whereas the choice of model is usually not. In Chapters 4 and 5 we shall investigate the concept of E-sufficiency in various settings with censoring problems and nuisance parameters. As has been suggested above, problems of censoring involve projections of functions into an appropriate subspace of functions as dictated by the censoring. The relevance of orthogonality and projection will become more apparent in the theory developed in Chapter 2. The problems of nuisance parameters have arisen where inference is required about one parameter θ in the absence of knowledge of parameter η, which can be a vector or scalar parameter. In extreme cases, such as the Neyman-Scott paradox, the m.l.e. $\hat{\theta}$ becomes inconsistent. However the concept of unbiasedness of inference functions provides clues to the difficulty. Define the profile log-likelihood to be $l_{PR}(\theta) = l(\theta, \hat{\eta}(\theta))$ where $\hat{\eta}(\theta)$ is the vector m.l.e. of nuisance parameters for given θ. The profile score function then becomes $S_{PR}(\theta) = \frac{\partial}{\partial \theta} l_{PR}(\theta)$. However, it is easily seen that the profile score function is not unbiased. For the Neyman-Scott paradox, this bias is strong enough to destroy consistency. A way around this difficulty is to construct an unbiased function such as

$$\psi(\theta) = S_{PR}(\theta) - E_\theta S_{PR}(\theta). \tag{1.8}$$

Of course the expectation above will depend in general upon more than just θ. This can be handled by various methods, but we will postpone this discussion until Chapter

4. Another problem with nuisance parameters arises when confidence statements about θ depend upon the unknown nuisance parameters.

In Chapter 6 the general theory developed in Chapter 2 is applied to inference problems involving parameters of stochastic processes. This area merits particular attention because the traditional likelihood theory becomes difficult to apply in practice for models of stochastic processes. It will be seen that by suitably restricting the class of methods available for inference it is possible to obtain a theory of inference that avoids many of the problems of maximum likelihood. We will leave the detailed discussion of this until Chapter 6. Traditional maximum likelihood has features for some processes that are undesirable: difficulty in computation and lack of stability of the solution under perturbations of the data being two of them. Our approach to this will underline the prevailing philosophy of this monograph, which is to impose regularity on the class of allowable procedures, and then to search for an appropriate inference within that class.

Finally we note that the term "estimating function" is used by Godambe and others for functions ψ of both the data and the parameter. Nevertheless, these functions can be regarded as vehicles for more general focus of inference than simple point estimation. For this reason, we call these functions "inference functions" which we believe indicates their broad applicability to statistical inference.

Historical Development. We conclude this chapter with a description of the early development of estimating equations (or inference functions in our terminology). Any attempt to discuss the early mathematical ideas underlying the subject would take us too far afield. So we shall be content here to explain some work of a methodological nature.

The foundations of estimation started with early methods that provided a coherent framework for constructing estimators in a wide variety of settings. K. Pearson (1894) introduced the method of moments and applied it with some success to the problem of estimating parameters in a mixture of normal distributions. This is an example of a case where a set of estimating equations can be explicitly displayed but where the roots of these equations cannot be found explicitly. Fisher (1922) suggested the method of maximum likelihood as an alternative to the method of moments. From the present perspective, we can say that maximum likelihood has proved by far the more successful of the two (if popularity is anything to go by). The asymptotic inefficiency of method of moment estimators in many problems is a clear inferiority of the procedure to maximum likelihood. In small sample situations there is no obvious domination of one method uniformly over the other. It is interesting to note that the unbiasedness of the score function provides a link between the two methods. By solving the equation $S(\hat{\theta}) = 0$, we are essentially choosing that value of θ which matches $S(\theta)$ with its expectation.

The term "estimating function" was used in a distinctive early paper by Kimball (1946). In that paper, the concept of a stable estimating function was introduced as one whose expectation is constant in the parameter. Subsequently, Kendall (1951) used the term "unbiased" in connection with estimating functions, although the concept is not very different from that of stability as used by Kimball. Kimball also argued that the concept of sufficiency for statistics is not adequate for estimating functions. While our approach to sufficiency in the next chapter is rather different from that of Kimball (1946), we agree with the broad principle that sufficiency for estimating functions should be developed in its own right.

In tracing the early work on the concept of an efficient estimating equation, we need to distinguish between its use as a tool for asymptotic inference and its use as a criterion in its own right. For example, in Wilks (1938), the idea is used to discuss asymptotic confidence intervals. In Durbin (1960), the optimality of a linear unbiased estimating equation was introduced. By contrast with Wilks (1938), we note the use of optimality on the equation itself, and not just for the purpose of deriving efficient estimates. Godambe (1960) suggested a more general definition while at about the same time, G. Barnard, in a private communication, recommended this generalization to Durbin. Both Barnard and Godambe noticed the optimality of the score function under this criterion. We shall not discuss this optimality criterion *per se* in the pages that follow as it can be seen that under fairly general circumstances the notion of an "optimal" function is equivalent to a first order E-sufficient function as defined in Chapter 2.

It is impossible to do full justice here to the proliferation of papers in the applications of estimating equations since 1960. The basic concepts of Chapter 2 can be found in Small and McLeish (1988). Of a theoretical nature, papers by Kale (1962) and Godambe and Thompson (1978) are of interest. More recent work also includes Godambe (1984), Kale (1986) and McLeish (1983,1984). We shall return to these and other papers in subsequent chapters.

CHAPTER 2
THE SPACE OF INFERENCE FUNCTIONS:
ANCILLARITY, SUFFICIENCY AND PROJECTION

2.1 Basic Definitions.

In this chapter, we construct the space of inference functions and information-theoretic notions of E-sufficiency and E-ancillarity within this space. Let $\mathcal{X}$ be a sample space, and $\mathcal{P}$ be a class of probability measures P on $\mathcal{X}$. For each $P\epsilon\mathcal{P}$ we let V_P be the vector space of real valued functions f defined on the sample space $\mathcal{X}$ such that $E_P[f(X)]^2 < \infty$. We introduce the usual inner product defined on V_P ,

$$<f_1,f_2>_P = E_P\{f_1(X)f_2(X)\}$$

Let θ be a real valued function on the class of probability measures $\mathcal{P}$ and define the parameter space $\Theta = \{\theta(P); P\epsilon\mathcal{P}\}$. Note that θ need not be a one to one functional. If it is, we call the model a *one-parameter model*.

The fundamental objects of our analysis will be *inference functions* i.e. functions $\psi:\Theta \rightarrow \bigcup_P V_P$ such that $\psi(\theta(P)) \epsilon V_P$ for all $P\epsilon\mathcal{P}$. The function ψ is said to be *unbiased* if $<\psi(\theta(P)), 1>_P=0$ for all $P\epsilon\mathcal{P}$. Two functions ψ and ϕ are said to have *constant covariant structure* if for all $P\epsilon\mathcal{P}$ the inner product

$$<\psi(\theta(P)),\phi(\theta(P))>_P$$

is a function of P only through $\theta = \theta(P)$. In this circumstance we write $<\psi,\phi>_{\theta}$ for this quantity as a function of θ. Let Ψ be a set of unbiased inference functions such that any pair of functions from Ψ have constant covariant structure. Later we will construct various examples of such spaces. Clearly Ψ can be made into a vector space by closing it under pointwise addition: $(\psi_1 + \psi_2)(\theta) = \psi_1(\theta) + \psi_2(\theta)$, and multiplication:

$(k\psi)(\theta) = k\psi(\theta)$. More generally, we can allow k to be a non-random function of $\theta \epsilon \Theta$.

Henceforth, by a "vector space" or "linear space" of inference functions we shall mean a

vector space in this more general sense: the space will be required to be closed under

multiplication by scalars $k(\theta)$ depending on the parameter but not on the data. So Ψ

is endowed with a family of inner products $<\psi_1,\psi_2>_\theta$ for every $\theta \epsilon \Theta$. We now introduce

a topology on Ψ. Let Ψ_θ be the set of functions $\psi(\theta)$ for $\psi \epsilon \Psi$. The evaluation func-

tion $e_\theta : \Psi \to \Psi_\theta$ for which $e_\theta(\psi) = \psi(\theta)$ is a function into a space of square integrable func-

tions. For each $\theta \epsilon \Theta$ we can endow Ψ_θ with the natural norm $\|v\|_\theta^2 = <v,v>_\theta$. Call

the associated topology the θ-topology of V_θ. We define the *weak square* topology of

Ψ as the weakest topology making the evaluation function e_θ continuous for every $\theta \epsilon \Theta$

when Ψ_θ has its θ-topology. Convergence in this topology is characterized as follows:

$\psi_n \to \psi$ if and only if $<\psi_n - \psi, \psi_n - \psi>_\theta \to 0$ for all $\theta \epsilon \Theta$. For the purposes of subsequent

analysis, an additional condition is imposed upon Ψ, namely that Ψ_θ is a Hilbert

space with respect to the inner product above. This requires that if ψ_n is a sequence

of inference functions in Ψ such that the double limit $\lim_{nm} \|\psi_n - \psi_m\|_\theta$ equals zero,

then there exists a function $\psi \epsilon \Psi$ such that $\lim_n \|\psi_n - \psi\|_\theta = 0$. When Ψ_θ is a Hilbert

space for all $\theta \epsilon \Theta$ we say that the space Ψ is a Hilbert space of inference functions.

Note that we are using the concept of a Hilbert space in a more relaxed sense than

usual. The weak-square norm defined on Ψ_θ is really a semi-norm in general. We

shall only require that the inner product of a Hilbert space be positive semi-definite.

The conditions on Ψ can be summarized by stating that

henceforth the inference function space Ψ shall be assumed to be a Hilbert space of *unbiased functions, any pair of which have constant covariant structure.*

An ancillary statistic is one whose distribution is insensitive to changes in the parameter. However, our window on the distribution is the inference function ψ and we assume that changes in the parameter are first evident through changes in the expectation of the inference function. From this perspective, the following definition is a natural one.

2.1.1 Definition. An unbiased inference function $\phi \epsilon \Psi$ is said to be *E-ancillary* if ϕ can be written as the weak square limit of functions ψ such that

$$E_Q \psi(\theta(P)) = 0$$

for all $Q \epsilon \mathcal{P}$. We let $\mathcal{A}$ be the set of E-ancillary functions in Ψ. By construction it is closed and it is easy to see that it is a linear subspace of Ψ.

2.1.2 Example. Let $\mathcal{P}$ be a 1-parameter model of unbiased inference functions. Suppose T is an ancillary statistic in the usual sense that its distribution does not depend on the parameter $\theta \epsilon \Theta$. Then for any values of the parameters θ, η and inference function $\psi \epsilon \Psi$, if $\psi(\theta)$ is $\sigma(T)$ measurable for each θ,

$$E_\eta[\psi(\theta)] = E_\theta[\psi(\theta)] = 0.$$

Thus the inference functions in Ψ that are functions of an ancillary statistic form an E-ancillary class in the sense that they are in $\mathcal{A}$.

In a sense the sufficient statistics contain the orthogonal complement to the set of ancillary statistics, at least under some conditions (see for example Basu's theorem). This motivates the following extension of the notion of sufficiency.

2.1.3 Definition. Let S be a subset of Ψ. We say that S is an *E-sufficient* subset of inference functions if the condition that $<\psi,\phi>_{\theta} = 0$ for all $\theta \epsilon \Theta$ and for all $\psi \epsilon S$ implies that ϕ is an E-ancillary function.

2.1.4 Example. Again we assume a 1-parameter model. Let Ψ be the space of *all* unbiased inference functions possessing finite second moments. Suppose T is a sufficient statistic for θ. Let S be the space of all $\sigma(T)$-measurable inference functions in Ψ. Suppose ϕ is any element of Ψ such that $<\phi,\psi>_{\theta} = 0$ for all $\theta \epsilon \Theta$ and for all $\psi \epsilon S$. It will be shown that $\phi \epsilon \mathcal{A}$, and therefore that S is E-sufficient. For each $\theta \epsilon \Theta$, let $supp(\theta) = supp(\theta(P))$ be the support of the distribution of P. Observe that the sufficiency of T requires that $supp(\theta)$ be a $\sigma(T)$ measurable set. Let I_{θ} denote the indicator function of the support of θ. Then $<\psi,\phi I_{\theta}>_{\theta} = <\psi,\phi>_{\theta} = 0$ for all θ and for all $\psi \epsilon S$. Setting $\psi(\theta) = E_{\theta}[\phi(\theta)I_{\theta}|T]$, we see that

$$E_{\theta}[\phi(\theta)|T] = 0 \quad a.s. \quad on \quad supp(\theta). \tag{2.1}$$

However, sufficiency of T implies that

$$0 = E_{\theta}[\phi(\theta)|T] = E_{\eta}[\phi(\theta)|T] \quad a.s. \quad on \quad supp(\theta) \bigcap supp(\eta).$$

Thus $E_{\eta}[\phi(\theta)I_{\theta}] = E_{\eta}[\phi(\theta)I_{\theta}I_{\eta}] = E_{\eta}\{I_{\eta}E_{\eta}[\phi(\theta)I_{\theta}|T]\} = E_{\eta}\{I_{\theta}I_{\eta}E_{\theta}[\phi(\theta)|T]\} = 0$, and therefore $\phi(\theta)I_{\theta}$ is E-ancillary. But $\mathcal{A}$ is weak-square closed, and $||\phi-\phi I_{\theta}||_{\theta}=0$. We conclude that $\phi \epsilon \mathcal{A}$.

The reader should notice that the proof depends upon the fact that the conditional expectation of an unbiased inference function is unbiased, and therefore is an element of Ψ. When Ψ is a more restricted space, the space of $\sigma(T)$-measurable functions can even be empty and not E-sufficient. However, we shall see that E-sufficient subspaces still exist and remain appropriate classes for inference in the restricted setting.

In a sense, a complete sufficient statistic forms the orthogonal complement of the space of ancillary statistics. This motivates the definition:

2.1.5 Definition. A subset S of Ψ is said to be a *complete* E-sufficient subset if (A) and (B) below are equivalent for all $\phi\epsilon\Psi$.

(A) $<\psi,\phi>_\theta = 0$ for all $\theta\epsilon\Theta$, $\psi\epsilon$ S

(B) $\phi\epsilon$ $\mathcal{A}$

Ignoring, for the moment, the question of whether a complete E-sufficient space exists, we first note that if it does exist, it must be uniquely defined and closed in the weak-square topology. In fact, we shall see that complete E-sufficient spaces exist in abundance. Henceforth S shall denote the complete E-sufficient subset. It should also be noted that the subsets $\mathcal{A}$ and S are linear subspaces of Ψ.

2.1.6 Example. Again we consider a 1-parameter model, and let Ψ be the space of all unbiased inference functions. Suppose T is a complete sufficient statistic. Let S be the space of $\sigma(T)$-measurable functions in Ψ. Then (A) implies (B) above. Conversely, suppose $E_\eta\phi(\theta;X)=0$ for all η. Define

$$g(t)=E_\eta[\phi(\theta;X)|T=t] \quad \text{for} \quad (T=t)\subset supp(\eta). \tag{2.2}$$

It must first be proved that the function g is well defined through (2.2) on the set $\bigcup_\eta \{t:(T=t)\subset supp(\eta)\}$. As in Example 2.1.4, the sufficiency of T implies that I_η is $\sigma(T)$ measurable for all values of η. This implies that either $(T=t)\subset supp(\eta)$ or $(T=t)\cap supp(\eta)$ is the empty set. Suppose η and ξ are two parameter values for which $(T=t)\subset supp(\eta)\cap supp(\xi)$. Then the sufficiency of T ensures that $E_\eta[\phi(\theta)|T=t]=E_\xi[\phi(\theta)|T=t]$. Arguing in this fashion we see that g is well defined. Now

$$E_\eta[g(T)] = E_\eta[\phi(\theta)] = 0 \tag{2.3}$$

for all η. As T is also complete, it follows that $g(T)=0$ $\eta-a.s.$ for all η. Suppose $\psi\epsilon S$ and $\theta\epsilon\Theta$. Then

$$<\psi,\phi>_\theta=E_\theta\{E_\theta[\psi(\theta)\phi(\theta)|T]\}=E_\theta[\psi(\theta)g(T)]=0. \tag{2.4}$$

By extension to the closure of the set of such ϕ we see that $<\psi,\phi>_\theta=0$ for all $\phi\epsilon\mathcal{A}$.

The example above establishes the existence of a complete E-sufficient subspace for 1-parameter models possessing a complete sufficient statistic. However, it will be shown in the next section that complete E-sufficient spaces will exist in many cases even when the model does not have a complete sufficient statistic for the parameter.

2.2 Projections and Product Sets.

Suppose an inference function ψ has been found to estimate $\theta\epsilon\Theta$. For a variety of reasons, ψ may not be an appropriate estimating function for the problem because it may depend on unobserved data, or not lie in the E-sufficient subspace, or it may lack computational simplicity or robustness as insurance against misspecification of the model. In general, we might suppose that there is some subset Υ in Ψ which contains the candidate inference functions for the problem. Then we would wish to replace the function ψ with some $\tilde{\psi}\epsilon\Upsilon$.

The method we suggest here is to choose $\tilde{\psi}$ to be an element of Υ that is closest in a sense to be defined to the original inference function ψ. We can write

$$||\psi-\tilde{\psi}||_\theta^2 = <\psi-\tilde{\psi},\psi-\tilde{\psi}>_\theta$$

It seems reasonable to define $\tilde{\psi}(\theta)$ pointwise in θ to be that value which minimizes $||\psi-\tilde{\psi}||_\theta^2$ for all $\tilde{\psi}\epsilon\Upsilon$. While we can define a function $\tilde{\psi}$ pointwise in this way, there is no guarantee that this function will lie in Υ: i.e. there will not always be a $\tilde{\psi}\epsilon\Upsilon$ which uniformly minimizes the distance for all θ. However, we shall see that there are classes of sets Υ for which the pointwise minimizing $\tilde{\psi}$ remains in the set.

2.2.1 Definition. Let $\Upsilon\subset\Psi$. We set $\Upsilon_\theta = \{\phi(\theta);\phi\epsilon\Upsilon\}$. The set Υ is said to be a *product set* if $\bigcap_{\theta\epsilon\Theta}\{\psi;\psi(\theta)\epsilon\Upsilon_\theta\} = \Upsilon$.

A product set is also called a *box set* in the literature. Then the standard result follows:

2.2.2 Proposition. The intersection of product sets is a product set.

When Υ is a product set we can also write $\Upsilon = \underset{\iota \in \Theta}{\times} \Upsilon_\iota$. The weak-squared topology is the weakest topology having all coordinate projections continuous. It is well known that

$$Cl(\Upsilon) = Cl(\underset{\iota \in \Theta}{\times} \Upsilon_\iota) = \underset{\iota \in \Theta}{\times} Cl(\Upsilon_\iota) \tag{2.5}$$

where Cl denotes the appropriate closure in each setting. From this and the definition of the ancillary subspace,

2.2.3 Proposition. The weak-square closure of a product set is a product set.

2.2.4 Corollary. Suppose Ψ is a product set. Then $\mathcal{A}$ is a product set.

The following proposition demonstrates that $\mathcal{S}$ is also a product space.

2.2.5 Proposition. Suppose Ψ is a product set. Then there exists a complete E-sufficient space.

Proof. Let $\mathcal{S}$ be the space of all functions $\psi \epsilon \Psi$ such that $<\psi, \phi>_\iota = 0$ for every $\theta \epsilon \Theta$ and every $\phi \epsilon \mathcal{A}$. It can be seen that the space $\mathcal{S}$ is non-empty because the zero-function is in $\mathcal{S}$. As defined, $\mathcal{S}$ is a closed linear product set such that $\mathcal{S}_\iota$ is the orthogonal complement of $\mathcal{A}_\iota$ in Ψ_ι. That (A) and (B) are equivalent in the definition of complete E-sufficiency now follows.

The following proposition now tells us that we can find an element of a closed product set which is closest to a given inference function in Ψ.

2.2.6 Proposition. Let Υ be a closed product set in Ψ. Let $\psi \epsilon \Psi$. Then there exists a $\tilde{\psi} \epsilon \Upsilon$ such that for every $\theta \epsilon \Theta$,

$$||\psi - \tilde{\psi}||_{\tilde{\theta}}^2 = \inf_{\phi \epsilon \Upsilon} ||\psi - \phi||_{\tilde{\theta}}^2 \tag{2.6}$$

Proof. For each $\theta \epsilon \Theta$, let $\tilde{\psi}' \epsilon \Psi$ be chosen so that $\tilde{\psi}'(\theta)$ lies in Ψ_θ and minimizes $E_\theta\{[\psi(\theta) - v]^2\}$ among all $v \epsilon \Upsilon_\theta$. As Υ is closed, $\tilde{\psi}'(\theta)$ can be chosen to lie in Υ_θ. The inference function $\tilde{\psi}$ defined by

$$\tilde{\psi}(\theta) = \tilde{\psi}'(\theta) \tag{2.7}$$

will then be in Υ because Υ is a product set.

2.2.7 Proposition. Suppose Υ is a linear subspace of Ψ that is a closed product set. Let $P_1 \ll P_2$ for all P_1 and P_2. Then $\tilde{\psi}$ is P-a.s. unique for all P.

Proof. As Υ_θ is closed and convex in V_θ, it follows that $\tilde{\psi}'(\theta)$ is θ-a.s. unique for each $\theta \epsilon \Theta$. Absolute continuity of P_1 and P_2 now implies that $\tilde{\psi}'(\theta)$ is P-a.s. unique for all P.

In view of the above, we conclude this section with three common examples of linear subspaces that are closed product sets.

2.2.8 Example. Let Υ be the set of functions in Ψ that are $\sigma(T)$-measurable, where T is a measurable function on the sample space $\mathcal{X}$. Then provided Ψ is a product set it follows that Υ is a closed linear product space.

2.2.9 Example. Suppose $X = (X_1, X_2, \cdots, X_n)$ is a vector of random variables. Let Υ be the set of all functions in Ψ which are linear in the data for each value of $\theta \epsilon \Theta$. Then if Ψ is a product set it follows that Υ is a closed linear product space. Note that we can allow the coefficients of the linear functions to depend themselves non-linearly upon the parameter θ.

2.2.10 Example. Suppose $X_1, X_2, \cdots, X_n$ are i.i.d. random variables from a continuous distribution that is symmetrical about θ and with finite second moment. Let $\mathcal{W}$ be a linear space of vectors $(w_1(\theta), \cdots, w_n(\theta))$ which do not depend on the data and which satisfy the symmetry property that $w_i = w_{n+1-i}$ for all $i = 1,...,n$. We then define Υ to be the space of functions ψ of the form

$$\psi(\theta) = \sum_{i=1}^{n} w_i(X_{(i)} - \theta) \tag{2.8}$$

where $X_{(i)}$ is the $i-th$ order statistic, and $(w_1, \cdots, w_n)$ lies in $\mathcal{W}$. It can be seen that if the weights on the outer order statistics are chosen to be zero, this has the effect of trimming the data. The space Υ can be recognized as a space of inference functions based upon L-estimators of θ. Note that negative weights are permitted as seems to be appropriate for certain optimal L-estimators of heavy tailed distributions.

2.3 Ancillarity, Sufficiency and Projection for the 1-Parameter Model.

In this section we shall assume that we are working with a 1-parameter model in which all distributions are absolutely continuous with respect to each other so that likelihood ratios exist. In this setting the condition that Ψ have constant covariance structure is trivially satisfied for any space of unbiased inference functions having finite second moments with respect to all distributions in the model. Let Ψ be the set of all such inference functions. It is easy to see that Ψ will be a fortiori a Hilbert space. Let λ be a reference measure such that $P<<\lambda$ and $\lambda<<P$ for all $P\epsilon\mathcal{P}$. We define the likelihood

$$L(\theta) = dP/d\lambda \qquad (2.9)$$

where $\theta = \theta(P)$. Finally, we shall let $\mathcal{A}$ and $\mathcal{S}$ denote the space of E-ancillary functions and the complete E-sufficient space respectively. Both spaces are closed linear product subspaces of Ψ. Projections into $\mathcal{A}$ and $\mathcal{S}$ will exist and be λ-almost surely unique. If ψ is any element of Ψ then there is a λ-almost sure decomposition of ψ into $\psi = \psi_a + \psi_s$ where $\psi_a\epsilon\mathcal{A}$ and $\psi_s\epsilon\mathcal{S}$.

We now have the following result.

2.3.1 Proposition. The space $\mathcal{S}$ contains the weak-square closure of the class of functions $\psi\epsilon\Psi$ which can be represented in the form

$$\psi(\theta) = \int_\Theta [\frac{L(\eta)}{L(\theta)} - 1] \, d\Lambda_\theta(\eta) \qquad (2.10)$$

where, for each θ, Λ_θ is a signed measure of finite support on Θ.

Proof. Suppose $\phi \epsilon \mathcal{A}$. Then there exists a sequence ϕ_n converging to ϕ in Ψ in the weak square topology such that $E_\eta \phi_n(\theta) = 0$ for all n, θ and η. Then for any ψ with integral representation as above

$$< \psi, \phi >_{,} = \lim_n < \psi, \phi_n >_{,} \tag{2.11}$$

$$= \lim_n E_{,} \int_\Theta \phi_n(\theta)[\frac{L(\eta)}{L(\theta)} - 1] d\Lambda_{,}(\eta)$$

$$= \lim_n \int_\Theta E_\eta[\phi_n(\theta)] d\Lambda_{,}(\eta)$$

$$= 0.$$

From this it follows that $\psi \epsilon \mathcal{S}$. Because $\mathcal{S}$ is closed in the weak-square topology it also follows that any function that can be represented as the weak-square limit of a sequence of such functions will also lie in $\mathcal{S}$.

While 2.3.1 treats only the case of signed measures $\Lambda_{,}$ with finite support, it will typically be the case that a variety of inference functions lying in the complete E-sufficient subspace can be represented in the integral form of (2.10) without the finite support condition. We shall have use for such an example below.

2.3.2 Example. Suppose $X = (X_1, X_2, \cdots, X_n)$ is from a location model in the sense that X has joint p.d.f. $f(x_1 - \theta, \cdots, x_n - \theta)$. Let λ be Lebesgue measure on the real numbers. Define $\Lambda_{,}$ so that $\frac{d\Lambda_{,}}{d\lambda}(\eta) = \eta - \theta$ for $-M \leq \eta - \theta \leq +M$ and let $\psi(\theta)$ be defined through its integral representation as above (again, assuming that ψ lies in Ψ). We observe that $\psi(\hat{\theta}) = 0$ if

$$\hat{\theta} = \frac{\int\limits_{\theta-M}^{\theta+M} \eta \, f(x_1-\eta,\, \cdots ,x_n-\eta)d\eta}{\int\limits_{\theta-M}^{\theta+M} f(x_1-\eta,\, \cdots ,x_n-\eta)d\eta}. \tag{2.12}$$

So $\hat{\theta}$ is recognizable as a approximation to the Pitman estimator of θ. Letting $M \to \infty$ the Pitman estimator is obtained. We note that many minimum risk equivariant estimators of location can be obtained through similar arguments. More generally, the same is true of limiting Bayes estimators.

In view of the importance of the score function in the theory of unbiased functions (and indeed the importance of unbiasedness in the proof of the consistency of maximum likelihood estimation) it is natural to ask whether the score function $S(\theta)$ is an element of S. However, if the question is posed in total generality, then it can be seen that $S(\theta)$ need not even be an element of Ψ. This will occur when the second moment of the score function is infinite. Such pathologies need not concern us unduly, because the regularity conditions needed to ensure the usual asymptotic efficiency of the m.l.e. are in fact much more stringent than the requirement that $S(\theta)$ lie in Ψ. To see whether $S(\theta)$ is an element of S, consider the sequence of functions

$$\psi_n(\theta) = n\left[\frac{L(\theta+n^{-1})}{L(\theta)}-1\right] \qquad \text{for all } n \geq 1. \tag{2.13}$$

Suppose the likelihood function is differentiable. Provided that for n sufficiently large, $\psi_n(\theta)$ lies in the space S and is a Cauchy sequence in the weak-square metric for every θ, it will follow from the closure of S that the limit, namely $S(\theta)$ will be in the space S. The proposition above ensures that ψ_n lies in S provided that $\psi_n \epsilon \Psi$. In the next section we will discuss the role of the score function in greater generality than the

1-parameter model considered above.

Before leaving the study of complete E-sufficiency for the 1-parameter model we will discuss its relation to the classical notion of minimal sufficiency. It is apparent that among all E-sufficient spaces that are closed linear product subspaces of Ψ, the space S is the smallest in the sense that it is contained in all other such E-sufficient spaces. In this sense it is minimal. The concept of complete E-sufficiency is also reminiscent of minimal sufficiency in the fact that a complete E-sufficient space can be constructed in many cases where a minimal sufficient statistic exists but a complete sufficient statistic does not. In view of this it might be tempting to call the complete E-sufficient subspace a minimal E-sufficient subspace but in fact the projection into the complete E-sufficient subspace is stronger than a reduction to a minimal sufficient statistic. The closed linear space of unbiased functions which are measurable with respect to a minimal sufficient statistic T is E-sufficient and therefore contains the complete E-sufficient subspace. However, unless the minimal sufficient statistic is complete sufficient this containment will be strict. To see this, suppose a function of a minimal sufficient statistic is an unbiased estimator of zero. It then defines an E-ancillary function that lies within the space of $\sigma(T)$-measurable inference functions. So it follows that this function is not in S. In examples in which a complete sufficient statistic does not exist, this projection of a given inference function into S is analytically more difficult because unbiased estimators of zero in the model can be difficult to characterize. We shall return to this problem and suggest an approximation in the next chapter. For the moment, however, it can be noted that the existence of a complete E-sufficient space for models in which there is no complete sufficient statistic illustrates the advantage of defining completeness within an inference function space.

2.4 Local Concepts of Ancillarity and Sufficiency.

One of the principal advantages of viewing the data through the expected value or, as its sample analog, the observed value of an inference function is the ease with which definitions extend to "local" analogs. Local definitions of first and second order ancillarity and sufficiency are available in the literature (c.f. McCullagh (1984), Cox (1980)) but are rather complicated both to define and to use. For example, local second order ancillarity is defined as a statistic A whose distribution at $\theta_0 + \frac{\delta}{n^{1/2}}$ differs from that at θ_0 by terms of the order $O(n^{-1})$. The set of functions of the data independent of all second order ancillary statistics is called *second order locally sufficient*. Because we concentrate primarily on the expected value of an inference function, there is a natural extension of ancillarity defined as follows:

2.4.1 Definition. An unbiased inference function $\phi(\theta)$ is said to be *locally (first order) E-ancillary* if it can be written as a weak square limit of functions $\psi \epsilon \Psi$ such that

$$E_P \psi \{\theta(P_0)\} = o(\theta(P) - \theta(P_0)) \text{ as } \theta(P) \to \theta(P_0). \tag{2.14}$$

2.4.2 Example. Consider a simple one-parameter model with a family of distinct probability density functions $\{f_\theta(x); \theta \epsilon R\}$. Then an unbiased inference function ψ is locally E-ancillary at θ if

$$\frac{\partial}{\partial \eta} \int \psi(\theta) f_\eta(x) dx \,|_{\eta=\theta} = 0. \tag{2.15}$$

Assume sufficient regularity that we can interchange integrals and derivatives above so that (2.15) becomes

$$E_{\theta}[\psi(\theta)S(\theta)] = 0 \qquad (2.16)$$

with the score function $S(\theta) = \dfrac{\partial}{\partial \theta} \log f_{\theta}(x)$. Alternatively this can be written as

$<\psi,S>_{\theta} = 0$. Thus the locally E-ancillary inference functions are those which are orthogonal to the score function. Assuming that the score function is square integrable, such a set is closed in the weak-squared topology by definition.

An alternative formulation of the condition in the above definition which is useful when the inference function rather than the model is regular is obtained by noting that

$$\frac{E_P \psi(\theta_0)}{\theta(P_0) - \theta(P)} = \int \frac{\psi(\theta_0) - \psi(\theta(P))}{\theta(P_0) - \theta(P)} dP \qquad (2.17)$$

and as long as the integrand in the second expression is uniformly integrable and converges to the derivative of ψ at θ_0, we have that the function ψ' is unbiased. Therefore, this is an alternative definition of local E-ancillarity. By analogy with the concept of E-sufficiency we have the next definition.

2.4.3 Definition. A subset L is said to be *locally E-sufficient* if any $\phi \epsilon \Psi$ such that $<\phi,\psi>_{\theta} = 0$ for all $\psi \epsilon L$ and $\theta \epsilon \Theta$ is locally E-ancillary. The subset L is said to be *complete locally E-sufficient* if $\psi \epsilon L$ if and only if $<\phi,\psi>_{\theta} = 0$ for all locally E-ancillary ϕ.

It is immediate from this definition that a complete locally E-sufficient subspace is unique and is a closed linear subspace of Ψ that is a product set.

An immediate consequence for the 1-parameter model of the remarks above is that in the presence of the usual regularity, the complete locally E-sufficient can be represented as the space of all functions of the form $k(\theta)S(\theta)$ in Ψ. As these functions are more or less equivalent for the purposes of inference we shall abuse the terminology somewhat and speak of the score function $S(\theta)$ as locally E-sufficient. Note that this argument is particular to the 1-parameter model and that the (vector-valued) score function is not locally E-sufficient in nuisance parameter settings. Even here, however, local E-sufficiency has some value as shall be seen in Chapters 4 and 5.

2.5 Second Order Ancillarity and Sufficiency.

The local concepts discussed above might also be called first order concepts, as first derivatives are used. The extensions to higher orders suggest themselves, and in particular, the second order methods will be discussed here. By extension we replace the order condition for local E-ancillarity above with

$$E_P\psi\{\theta(P_0)\} = o[(\theta(P)-\theta(P_0))^2] \ \ as \ \theta(P)\to\theta(P_0). \tag{2.18}$$

The space of second order E-sufficient functions is then defined analogously to be those functions orthogonal to the second order E-ancillary functions. For 1-parameter models with standard regularity properties the second order E-sufficient functions are of the form $k_1(\theta)S(\theta) + k_2(\theta)[S^2(\theta) - I(\theta)]$, where $I(\theta)$ is the observed information function. We can decompose the space of second order E-sufficient functions into multiples of the score function, namely $\mathcal{L}$, and its orthogonal complement in the second order E-sufficient space. But the orthogonal complement of $\mathcal{L}$ must be contained in the class of first order E-ancillary functions. Thus the complement of L in the second order space can be characterized as the class of *first order E-ancillary second order E-sufficient functions.* In view of the fact that ancillarity is normally an undesirable property of an inference function, we might wonder whether the space of first order E-ancillary second order E-sufficient functions has any value for inference. It will be seen that in fact it does. Anticipating the arguments which follow we introduce the following notation.

2.5.1 Notation. Let T be the space of first order E-ancillary second order E-sufficient functions of Ψ. The space of second order E-sufficient functions will be written as $L + T$, i.e., the vector sum of the spaces L and T.

In order to discuss the relevance of T, a digression into Neyman-Pearson hypothesis testing is necessary. In testing a simple hypothesis $H: \theta = \theta_0$ versus a one-sided alternative, a UMP test may well exist in standard problems. Typically for two-sided alternatives, such a test does not exist. However, a UMP unbiased test may sometimes be found. Here we use the word "unbiased" in the classical Neyman-Pearson sense of a test whose power function $pow(\theta)$ is minimized on the null hypothesis θ_0. Even in this case, however, a general theory cannot be developed because UMP unbiased tests do not always exist. One solution to this problem is to relax the unbiasedness condition to local unbiasedness and demand that the test chosen be locally most powerful within a neighborhood of θ_0. A critical region of size α is called locally most powerful unbiased if its power function has the property that

$$pow(\theta_0) = \alpha \tag{2.19}$$

$$\frac{d}{d\theta}pow(\theta_0) = 0 \quad (locally\ unbiased) \tag{2.20}$$

$$\frac{d^2}{d\theta^2}pow(\theta_0) \quad is\ maximized \tag{2.21}$$

Such tests can be constructed under fairly general circumstances. However, Chernoff(1951) has pointed out a serious limitation of this approach that can occur in special cases. Suppose X has distribution $N(\theta,1/(1+\theta))$ where $\theta>-1$, and we wish to test the hypothesis that $\theta = 0$. It can be shown that the locally most powerful unbiased test of size α has a critical region C_α of the form

$$C_\alpha = \{y: y^4+c_\alpha y^2+y+d_\alpha \geq 0\} \tag{2.22}$$

where $y = (x-1)/2$. Unfortunately the family of critical regions so obtained has the counter-intuitive property that it is not nested: there exist $\alpha_1<\alpha_2$ such that C_{α_1} is not a subset of C_{α_2}. For a sensible theory of hypothesis testing, this fact is disconcerting! Normally, when a test statistic t is adopted, critical regions of the form $\{x: t(x)>t_\alpha\}$ are used, ensuring that the critical regions are naturally nested. A way out of this problem can be found by noting that the test region given above defines a family of test statistics

$$t(x) = [(x-1)/2]^4 + c[(x-1)/2]^2 + [(x-1)/2] \tag{2.23}$$

from which one might well be chosen for constructing critical regions for all values of α. Then it can be shown that all test statistics $t(x)$ above can be represented as $\psi(0)$, where $\psi\epsilon L+ T$. This suggests that the problem can be reformulated. Let us chose a test statistic for the hypothesis $H: \theta = \theta_0$ of the form $\psi(\theta_0)$ where $\psi\epsilon\Psi$. The analog of

Neyman-Pearson local unbiasedness for inference functions is that ψ is locally E-ancillary so that

$$< \psi , S >_{\theta_0} = 0 \qquad (2.24)$$

Note that the concept of power is replaced by the expectation function. The analog to maximizing the second derivative of the power function is to require that ψ be second order E-sufficient. This analogy needs a few words of explanation. There does not exist an inference function ψ for which $\frac{d^2}{d\theta^2} E_\theta \psi(\theta_0)|_{\theta \to \theta_0}$ is maximized subject to the constraint that ψ is first order E-ancillary at θ_0. By requiring second order E-sufficiency we demand that the function ψ be locally most sensitive (in our two-sided sense here) about θ_0 to different values of the parameter. The resulting family of solutions will generate the space $\mathcal{T}$ which can be thought of as a class of two-sided test statistics.

For the problem of testing $\theta = \theta_0$ the decomposition of second order E-sufficient functions has special significance. For local one-sided alternatives, the score function space $\mathcal{L}$ is appropriate and for local two-sided alternatives, the other component $\mathcal{T}$ can be recommended. Note that we can write $\mathcal{T} = \{k\psi: k \; varying\}$ for some ψ and thus, as for $\mathcal{L}$, the particular element of $\mathcal{T}$ chosen is not important provided k is non-zero. Of course, if a critical region of the form $\psi(\theta_0) > t$ is to be used, then ψ should be chosen so that $\frac{d^2}{d\theta^2} E_\theta \psi(\theta_0)|_{\theta \to \theta_0} > 0$.

The space of two-sided test functions $\mathcal{T}$ will often yield tests equivalent to the generalized likelihood ratio test statistics. For example, if $X_1, \cdots, X_n$ are independent normal random variables with mean θ and unit variance, then $\mathcal{T}$ is generated by

$$\psi(\theta_0) = [\sum_{i=1}^{n} (x_i - \theta_0)]^2 - n \tag{2.25}$$

which leads to the same test regions as the generalized likelihood ratio for a two-sided alternative. However, the two procedures are not equivalent in general. For suppose that X has a Poisson distribution with mean θ. In this case, the space $\mathcal{T}$ is generated by

$$\psi(\theta_0) = (x - \theta)^2 - x. \tag{2.26}$$

2.6 Parametrization Invariance of Local Constructions.

The first and second order analyses given above are dependent on the use of derivatives. As the derivative of any function of θ will depend upon the coordinate system for the parameter space it might be initially supposed that the methods described above are not parametrization invariant. In this regard, we note for example that the score function is not parametrization invariant in the following sense. Let $\theta_1 = \eta(\theta)$ be a reparametrization of Θ, and let the score function in the θ-parametrization be $S(\theta)$. The function $S(\eta^{-1}(\theta_1))$ is the natural unbiased function in the θ_1-parametrization that is derived from the original score function by the transformation η. However, because the score function is not invariant, it is not itself the score function for the reparametrized model. It is useful to note, however, that the transformed score function still lies in $\mathcal{L}$ in the new parametrization because it is a multiple of the score function. We can summarize this property by stating that the space $\mathcal{L}$ *is* parametrization invariant. A similar argument holds for the space $\mathcal{T}$.

2.7 Background Development.

For results related to 2.3.1, see Bartlett (1982). The notion that the score function is locally sufficient is discussed in a related sense to our own in Efron (1982). The local sufficiency is understood by Efron to be achieved through the approximation of the model by a tangent exponential family (for which the score function is sufficient). The concept of an E-ancillary inference function is a generalization of the concept of a first order ancillary statistic as defined in Lehmann (1981). Note that Lehmann uses the term "first order" in a different sense than our use in this chapter. So henceforth we call these first order ancillary statistics E-ancillary.

One of the interesting features of constructing a complete E-sufficient space is that it exists more generally than a complete sufficient statistic. To examine this, we need to turn briefly to Basu's Theorem which motivates the extension. Basu (1955,1958) proved that a complete sufficient statistic is independent of an ancillary statistic. The construction of UMVUEs from complete sufficient statistics does not actually use all the properties of complete sufficiency. Lehmann and Scheffe (1950) and Rao (1952) show that UMVUEs must necessarily be uncorrelated with all E-ancillary statistics. The restriction of the search for a UMVUE to statistics uncorrelated with E-ancillary statistics is analogous to our restriction to the complete E-sufficient subspace. However there is an important distinction between the two approaches. Just as we have generalized the concept of completeness by requiring such inference functions to be uncorrelated with members of the space $\mathcal{A}$, so Lehmann (1981, section 4) suggests that complete *statistics* can be generalized to the statistics that are uncorrelated with E-ancillary statistics. However, Lehmann notes that this idea does not lead to a much more general theory because, in the absence of a complete sufficient statistic, there may be no such statistic. In contrast, we have found a variety of examples. The difference between the two situations arises because our functions are explicitly a function of the parameter. Lehmann's statistics are required to be uncorrelated with all E-ancillaries for all parameter values, but may not depend upon the parameter itself.

The notion of *local sufficiency*, used in a sense that is closer to Fisher's original concept of sufficiency than is local E-sufficiency, is discussed by Efron (1982). Local sufficiency, in Efron's sense, is related to local E-sufficiency as defined in this chapter. If a model is an exponential family, then $\hat{\theta}$ will be sufficient and if it is not an exponential family, then the model can usually be approximated by an exponential family locally about any θ_0. The fact that $\hat{\theta}$ is sufficient in the exponential family approximation is described by Efron as local sufficiency in the original model. Efron (1982) also serves as an excellent overview of the uses of maximum likelihood estimation: estimation, summarization, and their relation to decision-theoretic methods such as UMVU estimation. Sprott and Viveros-Aguilera (1984) provide a different overview of maximum likelihood estimation based upon the construction of approximate pivotals.

As we have avoided the problem encountered in Lehmann (1981) we might well consider whether his program for the construction of UMVUEs might be continued within our inference function setting. Along these lines, the concept of unbiasedness of an inference function is not analogous to unbiasedness of an estimator. However a complete sufficient statistic *defines a parametrization* by being a unique UMVUE of its expectation. Its expectation is then linear in that newly defined parameter. By analogy we shall see in the next chapter that an element of the complete E-sufficient space defines a parametrization through its expectation $E_t \psi(\theta_0)$ at a given value θ_0. A uniqueness result also follows.

CHAPTER 3
SELECTING AN INFERENCE FUNCTION FOR 1-PARAMETER MODELS

In this chapter we restrict attention to the 1-parameter model, and assume that all distributions have common support.

We shall also suppose that Ψ is unrestricted, in the sense that it consists of all unbiased inference functions.

The reduction of the choice of inference function to those within a sufficient subspace of Ψ still leaves considerable choice as to the appropriate function ψ for estimation. We now consider various criteria for the selection of ψ and some of the properties of the associated estimators. As we have seen, the score function usually lies within the complete E-sufficient subspace, and so in many cases, our search for an appropriate inference function need go no further. However, the score function is only justified as locally E-sufficient and can have problems when certain global properties are required. For example, suppose $X_1, \cdots, X_n$ are independent random variables having Cauchy distributions centered at θ. Then

$$\lim_{\eta \to \pm\infty} E_\eta S(\theta) = 0. \qquad (3.1)$$

This can be interpreted as stating that the score function is E-ancillary "at infinity" for a Cauchy model. The trade-off is evident: to require E-sufficiency locally around the value θ, we are forced to accept E-ancillarity for alternatives at infinity. A consequence of this is that the score function will in general have multiple roots. In the case of the Cauchy distribution, these additional roots that do not yield consistent estimates will not disappear as the sample size goes to infinity. One way to attack this problem would be to choose a function ψ such that $E_\eta\psi(\theta)$ has some appropriate regularity as a function of η for every value of θ. Our first approach to this is to require linearity in η.

3.1 Linearization of Inference Functions.

For functions from a linear 1-parameter exponential family with complete sufficient statistic T, one natural parametrization is the expected value of T. In this case, the score function can be written in the form $S(\theta) = c(\theta)[T-\theta]$. One of the useful features of this parametrization is that $E_\eta S(\theta)$ is then seen to be linear in η. The requirement of linearity is a simple type of regularity that avoids the E-ancillarity problem mentioned above. In this section we consider the problem of choosing ψ in the complete E-sufficient subspace so as to be linear in η. It is easy to see that if such a function exists, it must be unique (up to an arbitrary multiple that can depend on θ but not on the data). For suppose that ψ_1 and ψ_2 are two such functions both lying in S. Then we can write

$$E_\eta\psi_1(\theta) = k_1(\theta)[\eta - \theta] \qquad (3.2)$$

and

$$E_\eta\psi_2(\theta) = k_2(\theta)[\eta - \theta] \qquad (3.3)$$

It follows that $k_2\psi_1-k_1\psi_2$ is an element of A. But as a linear combination of elements of S it must itself be in S which requires that $k_2\psi_1-k_1\psi_2 = 0$. If the complete E-sufficient subspace is generated by a complete sufficient statistic we have seen that such a linearization is possible in the appropriate parametrization. However, the existence of such a linearized function for a given parametrization in a general setting needs to be considered.

Let us return now to the general case of a 1-parameter model that need not necessarily be of the exponential family form. Projection turns out to be a useful tool for this question because we need only find any linearized inference function and then project it into the complete E-sufficient subspace. For example, suppose $X_1, \cdots, X_n$ are i.i.d. random variables from a location model with density $f(x-\theta)$. If each X_i has mean θ and finite variance, then $\psi(\theta) = \bar{x} - \theta$ is an element of Ψ with the required linearity, although it will not always lie within the complete E-sufficient subspace. However we can write $\psi = \psi_s + \psi_a$, where $\psi_s \epsilon S$ and $\psi_a \epsilon \mathcal{A}$. Then ψ_s will also be linear in η because $E_\eta \psi(\theta) = E_\eta \psi_s(\theta)$ for all $\eta, \theta \epsilon \Theta$. Of course, if a complete sufficient statistic exists, then this construction is equivalent to conditioning $\bar{X}$ on the complete sufficient statistic, and so is Rao-Blackwellization of $\bar{X}$. The root of the resulting inference function will therefore be the unique UMVUE for θ. However, the procedure can be applied in cases where no complete sufficient statistic exists, provided the image of ψ under projection can be found.

To study projection into S in the case where a complete sufficient statistic does not exist, note first that we can project ψ into the space of inference functions that are measurable with respect to a minimal sufficient statistic. Minimal sufficient statistics exist under fairly general conditions and projection in this step amounts to conditioning $\bar{X}$ upon a minimal sufficient statistic. Now the E-ancillary component of this function can be seen to be an unbiased estimator of zero. Suppose we let T be the minimal sufficient statistic for the model. Then the $\sigma(T)$-measurable image of ψ

under projection will be

$$\psi_T(\theta) = E(X_1 | T) - \theta. \tag{3.4}$$

The next step is to write ψ_T as the sum of an E-sufficient component and an E-ancillary component.

3.1.1 Example. Suppose that $f(x) = (1/2)e^{-|x|}$. Let T be the vector of order statistics $X_{(1)}, \cdots , X_{(n)}$. Then T is minimal sufficient. So $\psi(\theta) = \overline{X} - \theta$ already lies in the E-sufficient space generated by the minimal sufficient statistic. Does it lie in the complete E-sufficient subspace? For $n=1$ we can write

$$X - \theta = \int_{-\infty}^{+\infty} [\frac{L(\eta)}{L(\theta)} - 1] d\Lambda_\theta(\eta) \tag{3.5}$$

where $d\Lambda_\theta(\eta) = exp(\theta - \eta)$ for $\eta \geq \theta$ and $d\Lambda_\theta = exp(\theta + \eta)$ for $\eta < \theta$. So ψ_1 lies within the complete E-sufficient subspace. In fact for $n=1$, $\overline{X} = X$ is the unique uniformly minimum variance unbiased estimator. However for $n > 2$, ψ is not in the E-sufficient subspace. To see this, let $n=3$ and let $\phi(\theta) = X_{(1)} - 2X_{(2)} + X_{(3)}$. Then ϕ is E-ancillary and positively correlated with ψ. Larger values of n are similar.

Secondarily to our stated purpose, this example gives some idea of the difficulty of searching for a UMVUE in models which are not exponential families. This is especially interesting in location models because with the exception of normal models and the logarithm of gamma variates, they are never exponential families. Bondesson (1975) has investigated the existence of UMVUEs for location models, and has had to use quite heavy analytical machinery to obtain some restrictive results. He was able to show that UMVUEs usually do not exist in such models. Fortunately, complete E-sufficient functions that are linear in η can be found in more generality. All that is necessary to prove their existence is the existence of a function, linear in η, which does not lie within the space $\mathcal{A}$ of E-ancillary inference functions. The argument in the example is capable of generalization which we summarize with earlier ideas in the following proposition. First we introduce a formal definition.

3.1.2 Definition. An inference function $\psi(\theta)$ is said to be *E-linear* if $E_\eta \psi(\theta)$ is a linear function in η for every value of θ.

3.1.3 Proposition. If there exists an E-linear inference function ψ lying within the complete E-sufficient subspace then it is the unique such function within that subspace up to a multiple $k(\theta)$. Furthermore, if T is a UMVUE for θ, then $\psi(\theta) = k(\theta)[T-\theta]$ for some choice of $k(\theta)$.

Proof. Uniqueness has already been demonstrated. As noted by Lehmann and Scheffe (1950) and Rao (1952), if T is a UMVUE, then it is uncorrelated with every E-ancillary function. Therefore, it must lie in $\mathcal{S}$. As T is unbiased, $T-\theta$ is E-linear, and from uniqueness, the result is proved. Note that the existence of an E-linear function in $\mathcal{S}$ does not imply the existence of a UMVUE.

This theorem can be illustrated by the following example in which there does not exist a complete sufficient statistic.

3.1.4 Example. Suppose $X_1, \cdots, X_n$ are an i.i.d. sample uniformly distributed on the interval $[\theta - 1/2, \theta + 1/2]$. Let $X_{(1)}$ and $X_{(n)}$ be the smallest and largest order statistics respectively. If T is defined to be the ordered pair $(X_{(1)}, X_{(n)})$ then T is minimal sufficient but not complete. To find the E-linear element of the complete E-sufficient subspace we need to find the E-ancillary functions which lie within the space of inference functions measurable with respect to T. These are the functions which are measurable with respect to $A = X_{(n)} - X_{(1)}$. If we define $\psi(\theta) = X_{(1)} + X_{(n)} - 2\theta$ then ψ can be seen to have constant regression on A. So it follows that $\psi(\theta)$ is uncorrelated with respect to any function of A. This shows that ψ lies within the complete E-sufficient subspace. It is easily seen to be E-linear.

The analytical difficulty of projection into the complete E-sufficient subspace when the model is does not admit a complete sufficient statistic or is not a location model can be seen. In the next section, we perform a linearization similar to the one above but only in a local sense of the second order properties of the expectation function. This will have the advantage of being easier to calculate for many models.

3.2 Adjustments to Reduce Curvature.

Consider a general unbiased inference function of the form

$$\psi(\theta) = a(X)\theta - b(X) \tag{3.6}$$

and suppose that ψ is E-linear. It follows from this that $E_\theta\, a(X)$ is a constant independent of θ. Consequently

$$\frac{d}{d\theta}E_{\theta}a(X) = \frac{d}{d\theta}E_{\theta}\psi' = 0. \tag{3.7}$$

Furthermore, it is clear from the linearity that $E_{\theta}\psi''(\theta) = 0$

Consider for a moment a general unbiased inference function $\psi(\theta)$. Differentiating twice the unbiasedness condition,

$$\frac{d^2}{d\theta^2}\int \psi(\theta)L(\theta)d\lambda. = A + 2B + C = 0. \tag{3.8}$$

where, for the above linear inference function, we have shown that

$$A = E_{\theta}\frac{d^2}{d\theta^2}\psi(\theta) = 0 \tag{I}$$

and

$$B = \frac{d}{d\eta}E_{\eta}\frac{d}{d\theta}\psi(\theta)\big|_{\eta=\theta} = 0. \tag{II}$$

from which the third condition will always follow,

$$C = \frac{d^2}{d\eta^2}E_{\eta}\psi(\theta)\big|_{\eta=\theta} = 0 \tag{III}$$

Thus, the three conditions I,II,III above, are naturally linked to inference functions ψ which are approximately linear both pointwise and in expectation in the parameter θ. When this approximate linearity is desirable, for example when we wish a one-step estimator based on a linear approximation that is close to the root, we may choose from the second order locally E-sufficient subspace of inference functions, of the form $k_1S + k_2(S^2-I)$ one which satisfies the conditions I-III.

In general, in order to find $k_1(\theta)$, $k_2(\theta)$ it is convenient to find $k_2(\theta)$ in terms of $k_1(\theta)$ using III and then to use condition II to construct a first order linear differential equation in $k_1(\theta)$.

3.2.1 Example. Let $X_1, X_2, ..., X_n$ be independent identically distributed variates with score function $S(\theta) = c(\theta)[\sum_{i=1}^{n} X_i - n\theta]$. Then there is a unique (up to multiplication by a factor constant in X, θ) function ψ in the second order E-sufficient subspace satisfying I-III, namely

$$\psi(\theta) = \sum_{i=1}^{n} X_i - n\theta . \tag{3.9}$$

Somewhat more generally than the above example, it is easy to see that $k_2(\theta)$ can be chosen equal to 0 in any model for which the unbiased inference functions spanning the second order E-sufficient subspace , $S(\theta)$ and $S^2(\theta) - I(\theta)$, are orthogonal, since in this case,

$$\frac{d^2}{d\eta^2} E_\eta k_1(\theta) S(\theta)|_{\eta = \theta} \;\; = \;\; < k_1 S , S^2 - I >_\theta \;\; = \;\; 0. \tag{3.10}$$

and we may choose

$$k_1(\theta) = \exp\{ -1/2 \int \frac{E_\theta[I'(\theta)]}{E_\theta I(\theta)} d\theta \}. \tag{3.11}$$

so that condition I holds with $\psi = k_1 S$.

Thus, under the orthogonality of the functions S and $S^2 - I$, there is a simple non-random multiple of the score function that satisfies the conditions I-III. While members of the one-parameter exponential family parametrized by the mean as in example 3.2.1 are clearly included among such functions, they are by no means the only distributions with these properties. The following is a standard *mixture* model which, we show, also admits a multiple of the score function satisfying conditions I-III.

3.2.2 Example. Let f and g be two probability density functions with common support and suppose we observe $X_1, X_2, \cdots, X_n$, independent identically distributed random variates from the mixture probability density $\theta f(x) + (1-\theta)g(x)$, where $0 \le \theta \le 1$ is the mixture parameter. In this case the score function and information function can be written $S(\theta) = \sum_{i=1}^{n} S_i(\theta)$ and $I(\theta) = \sum_{i=1}^{n} S_i^2(\theta)$ where

$$S_i(\theta) = \frac{f-g}{\theta f + (1-\theta)g}(X_i) \tag{3.12}$$

In this case, $E(S^3) = E(SI)$ and solving the equation (4...),

$$k_1(\theta) = [E_\theta I(\theta)]^{-1} \tag{3.13}$$

Thus, the suggested function is $J^{-1}(\theta)S(\theta)$, where J is the expected information. However, it should be remarked that even within the one-parameter exponential family, the linearized element of the second order E-sufficient subspace is not always just a multiple of the score function. The next example is typical and shows that even here, useful estimators may result.

3.2.3 Example. By way of comparison with the linear one-parameter exponential family, consider observations X_i ;$i = 1, 2, ..., n$ from an extreme value probability density function of the form

$$f_\theta(x) = \exp\{(x-\theta) - e^{(x-\theta)}\} \tag{3.14}$$

This particular model has the properties both of a location family and an exponential family probability density with complete sufficient statistic $T = \sum_{i=1}^{n} \exp\{X_i\}$. The maximum likelihood estimator for θ is $\hat{\theta}_{MLE} = \log[n^{-1}\sum_{i=1}^{n} \exp\{X_i\}]$. Solving the equations I-III, we find that the curvature adjusted member of the second order E-sufficient space of functions is

$$\psi(\theta) = (4n+4)\sum_{i=1}^{n} \exp\{X_i - \theta\} - [\sum_{i=1}^{n} \exp\{X_i - \theta\}]^2 - 3n(n+1) \qquad (3.15)$$

Only one of the roots of this function is asymptotically consistent, and this involves a correction or adjustment to the maximum likelihood estimator and this is the root

$$\hat{\theta} = \hat{\theta}_{MLE} + \log[\frac{n}{(2n+2)-(n^2+5n+4)^{1/2}}] \qquad (3.16)$$

The above curvature adjusted estimator $\hat{\theta}$ has bias $o(1/n)$ whereas the maximum likelihood estimator $\hat{\theta}$ has bias asymptotic to $\frac{1}{2n}$. Thus we can interpret the curvature adjustment as a method for linearizing the estimating function and in so doing, compensating for the bias of the maximum likehood estimator.

3.3 Reducing the Number of Roots.

The techniques of E-linearization and second order curvature adjustment have some bearing upon the uniqueness of roots of inference functions, but do not address the problem directly. In this section we construct a family of inference functions with a varying number of roots.

As we have seen with the Cauchy distribution, the local E-sufficiency of the score function, providing as it does, sensitivity to local changes in the parameter about a given value θ, has no guarantee of being sensitive to values $\theta + \epsilon$ when ϵ is not vanishingly small. If multiple roots occur, we might seek to vary the sensitivity of the inference function at various distances ϵ to observe the effect on the roots.

Suppose, for example, $L(\theta)$ is a likelihood function defined on $(-\infty, +\infty)$ that is continuously differentiable with a finite number of local maxima and minima. Suppose furthermore that $L(\theta)$ vanishes at infinity. Define

$$\psi_0(\theta) = S(\theta) \tag{3.17}$$

$$\psi_\epsilon(\theta) = \frac{L(\theta + \epsilon) - L(\theta - \epsilon)}{2\epsilon L(\theta)} \quad , \epsilon > 0. \tag{3.18}$$

For a given data set, $S(\theta)$ might have multiple roots, but as ϵ increases, the number of roots will eventually diminish. To see this, let a be the smallest value of $L(\theta)$ such that $\frac{d}{d\theta} L(\theta) = 0$. Now choose ϵ so that 2ϵ is greater than the diameter of the set $\{\theta : L(\theta) = a\}$. Then it will be shown that ψ_ϵ has a unique root. That it has a root is evident from the continuity of L and the fact that it vanishes at infinity. Suppose that θ_1 and θ_2 are roots. As $L(\theta_1 + \epsilon) = L(\theta_1 - \epsilon) < a$ we see that $\theta_1 - \epsilon$ is less that all local maxima and minima of L and $\theta_1 + \epsilon$ is greater than all such maxima and minima. The same is true for θ_2. If $\theta_2 > \theta_1$, then $L(\theta_2 - \epsilon) > L(\theta_1 - \epsilon)$, and $L(\theta_2 + \epsilon) < L(\theta_1 + \epsilon)$. But this cannot be. Similarly, we cannot have $\theta_1 > \theta_2$. So $\theta_1 = \theta_2$, and the result is proved.

Note that the number of roots of ψ_ϵ is not necessarily monotone decreasing in ϵ. However, as ϵ increases, the overall trend in the number of roots is to decrease. This root reduction property suggests a strategy for identifying inappropriate roots of the score function. By increasing ϵ, we eventually reduce the number of roots to one, or to a set in which there is a clear choice as to which is appropriate. This raises the question as to whether the root of ψ_ϵ is a good estimator when ϵ is not vanishingly small. It should be noted that ψ_ϵ lies within the complete E-sufficient space provided the first and second moment conditions are satisfied. Suppose the likelihood arises from a location model. Then $X_1-\theta, \cdots, X_n-\theta$ are i.i.d. pivotal quantities. Consider the problem of estimating θ by an interval estimate of the form $[\hat{\theta}-\epsilon, \hat{\theta}+\epsilon]$. Then $\hat{\theta}$ would be chosen so as to maximize the coverage probability. If we are restricted to location equivariant choices for $\hat{\theta}$, then the optimum $\hat{\theta}$ corresponds to a minimum risk equivariant estimator for the loss function

$$K(\hat{\theta},\theta) = 0 \ \ \text{for} \ \ |\hat{\theta}-\theta| \leq \epsilon \tag{3.19}$$

$$K(\hat{\theta},\theta) = 1 \ \ \text{for} \ \ |\hat{\theta}-\theta| > \epsilon. \tag{3.20}$$

We can formally identify the minimum risk equivariant estimator $\hat{\theta}$ with the Bayes estimator for the loss function K based upon the uniform improper prior on $(-\infty, +\infty)$. However, such an estimator can then be seen to be a solution to the equation $\psi_\epsilon(\hat{\theta}) = 0$. So for location models, the roots of ψ_ϵ are of some interest.

This argument suggests that we should modify ψ_ϵ for scale models to be of the form

$$\psi_1(\theta) = S(\theta) \tag{3.21}$$

$$\psi_\epsilon(\theta) = \frac{L(\epsilon\theta)-L(\epsilon^{-1}\theta)}{\theta L(\theta)[\epsilon-\epsilon^{-1}]} , \ \ \epsilon > 1. \tag{3.22}$$

Consider an example of four observations from a Cauchy distribution centered at zero, in which the score function has multiple roots. Here we have $x_1=6.512$, $x_2=-9.042$, $x_3=0.130$, $x_4=0.805$. As can be seen from figures 3.1,3.2 and 3.3, the number of roots of ψ_ϵ is 5,3, and 1 for $\epsilon =0.0$, 2.0 and 4.0 respectively.

The reader will have noted that for neighboring values of ϵ, the roots of ψ_ϵ can be considered to be perturbations of each other. This raises the question as to which root of the score function corresponds to the unique root of ψ_ϵ for large ϵ. To determine this, consider the set Q of all points (ϵ,θ) such that $\psi_\epsilon(\theta) = 0$, for $0\leq\epsilon\leq t$, where t is chosen so that ψ_t has a unique root $\hat{\theta}_t$. Suppose L has a unique global maximum. Then it can be shown that $(0,\hat{\theta})$ and $(t,\hat{\theta}_t)$ lie in the same connected component of Q if and only if $\hat{\theta}$ is the unique global maximum of L. Thus we see that the unique root of ψ_t can be regarded as a perturbed global maximum of the likelihood.

In the absence of Wald's conditions for the consistency of the global m.l.e., there might exist a consistent local maximum of the likelihood which is not the global maximum. This problem is typically attacked by selecting an $n^{1/2}-$ consistent estimator of θ and iterating via Newton's method from this estimate to a root of $S(\theta)$. However, while the resulting procedure is justifiable asymptotically, the property of $n^{1/2}-$ consistency is too weak to ensure sensible estimates on small samples. Curiously, the problem of using ψ_ϵ are just the opposite. By reducing the number of roots, ψ_ϵ provides more straightforward estimates for small samples. However, to guarantee consistency, ϵ must go to zero, which in turn can introduce multiple roots into the inference function. The fact that one method works on large samples and the other on small suggests that a combination of the two approaches might work well. Suppose $\tilde{\theta}$ is an $n^{1/2}$–consistent estimator which need not have desirable small sample properties. We would reasonably expect the error in this estimate to be about the same order as the square root of its mean square error. Suppose we set ϵ to be equal to the square root of the mean square error of $\tilde{\theta}$. We can then choose that root of ψ_ϵ found by iterating via Newton's method from $\tilde{\theta}$. This has the advantage that ψ_ϵ will be sensitive to departures from the true value of the parameter on the same order of magnitude as the uncertainty in $\tilde{\theta}$. For small samples, where $\tilde{\theta}$ is a poor estimate ψ_ϵ will likely have one root within the range of variation of $\tilde{\theta}$. For large samples and small ϵ, the problem of multiple roots is handled by the convergence of $\tilde{\theta}$ to the consistent root.

In the last two sections of this chapter we consider two related methods which are closely associated with fiducial inference and the construction of pivotal quantities. They represent a small departure from the general philosophy of this monograph that the quality of an inference function can be appropriately analyzed through its expectation. We will be concerned below with the distribution of $\psi(\theta)$ in the spirit of pivotal inference. As a result of this, the criteria for selection that are outlined below will not be preserved under projection into a closed linear product space.

3.4. Median Adjustment.

One of the most useful features of the score function $S(\theta)$ is that for moderately large sample sizes from models satisfying the usual regularity conditions, it is approximately normally distributed with mean 0 and variance $J(\theta) = -E_{\theta}S'(\theta)$. This permits the usual construction of confidence sets for θ of the form

$$\{\theta\,;|S(\theta)|\leq z_{\alpha/2}J^{1/2}(\theta)\} \tag{3.23}$$

The observed information sometimes replaces the expected information in the above construction, but in either case, the argument supporting this as a confidence set is essentially asymptotic. It might seem reasonable that if we wish such confidence statements to apply in smaller samples, we might wish to adjust the score function $S(\theta)$ by correcting for the skewness or otherwise attempting to obtain an inference function with greater regularity (more normal-like, for example) than the score function. For example. Cox and Hinkley (1974, page 341) adjust with second order terms setting

$$\psi = [S - k(S^2 - J)]J^{-1/2} \tag{3.24}$$

However, this has the undesirable feature of making the measure of discrepancy typically dependent on the 6'th moment of the score function, and we might question whether model assumptions are accurate at this level of delicacy. This skewness correction is order $n^{-1/2}$.

An alternative approach that we consider here involves the selection of inference function $\psi \epsilon S$ so as to correct for asymmetry in a cruder fashion. Suppose we choose $\psi \epsilon S$ so that the median and the mean coincide, or $med_{\theta}\psi(\theta) = E_{\theta}\psi(\theta) = 0$.

For example, if we restrict to the class of functions of the form

$$\psi(\theta) = \frac{L[\xi(\theta)]}{L(\theta)} - 1 \tag{3.25}$$

then we would seek $\xi(\theta)$ satisfying the above symmetrizing condition on the median. For example, suppose $X_1, X_2, \cdots, X_n$ are independent identically distributed exponential variates with mean θ. In this case, the score function has shifted gamma distribution and is quite asymmetrical for $n \leq 4$. However, for each value of θ, there exists a unique $\xi(\theta)$ such that the function $\psi(\theta)$ given above is median as well as mean unbiased. It can be shown that $\xi(\theta)$ is approximately $2^{-1/n}$ (exact for $n = 1$). For $n = 3$, for example, $\psi(\theta)$ has probability density function

$$g(t) = 1.978[log(1+t) - log2]^2(1+t)^{2.847} \tag{3.26}$$

which is close to being perfectly symmetrical about 0 and remarkably close in shape to the normal probability density function (see figure 3.4). We shall discuss this example with censorship in chapter 5.

Provided that the inference function $\psi(\theta)$ as defined above is a strictly monotone function of θ, it will also follow that $P_\theta[\hat{\theta} \geq \theta] = 1/2$ so that $\hat{\theta}$ is median unbiased.

A comparison can also be made with the score function under standard asymptotic conditions. Let $\hat{\theta}_{MLE}$ be the maximum likelihood estimate. For $\theta - \hat{\theta}_{MLE} = o(n^{-1/2})$ we can write

$$\psi(\theta) = -1/2I(\hat{\theta}_{MLE})[(\xi(\theta) - \hat{\theta}_{MLE})^2 - (\theta - \hat{\theta}_{MLE})^2] + o(1) \tag{3.27}$$

But $P_\theta[\psi(\theta) \geq 0] = 1/2$ requires that

$$[\hat{\theta}_{MLE} - \xi(\theta)]^2 - (\hat{\theta}_{MLE} - \theta)^2 = o(n^{-1}) \tag{3.28}$$

implying that $\xi(\theta) - \theta = o(n^{-1/2})$. But this together with

$$-\psi(\hat{\theta}) = I(\hat{\theta}_{MLE})[(\xi(\hat{\theta}) - \hat{\theta}_{MLE})(\frac{\xi(\hat{\theta}) + \hat{\theta}}{2} - \hat{\theta}_{MLE})] + o(1) = 0 \tag{3.29}$$

requires that $\hat{\theta}_{MLE} - \hat{\theta} = o(n^{-1/2})$. Therefore $\hat{\theta}$ is asymptotically efficient.

3.5. Approximate Normal Inference Functions.

Consider a one-parameter model with sufficient statistic T. Suppose we denote the cumulative distribution function of T by $F_\theta(t)$. Then there is a transformation of this sufficient statistic which is normally distributed and this function $\Phi^{-1}(F_\theta(T))$ is unique among standard normally distributed unbiased inference functions that are increasing functions of T. Therefore, simple transformations of T to approximate normality correspond to simple approximations of the inverse normal c.d.f. For example, the approximation

$$\Phi^{-1}(y) = \frac{\{-\ln(1-y)\}^{1/3} - \mu}{\sigma} \tag{3.30}$$

where μ is chosen so that the function is unbiased provides a remarkably accurate simple approximation to the normal inverse c.d.f. and corresponds to transforming an exponential variate T by the power transformation $T^{1/3}$. Figure 3.1 shows the accuracy of this approximation. What is especially remarkable about this case is that when the sufficient statistic T has a an arbitrary chi-squared distribution, the same transformation seems to work well (c.f. Sprott and Viveros) indicating that the function $\Phi^{-1}(F(T))$ is approximately a linear function of $T^{1/3}$ where F is a chi-squared c.d.f. with an arbitrary number of degrees of freedom.

The general question of whether there is a single transformation that will simultaneously transform variates from a parametric family to approximate normality is closely related to the question of which parametric families are close to normality for all values of the parameter. Let us consider this second question first. Consider a one-parameter model with probability density functions $f(x;\theta)$. One rough measure of approximate normality is the third derivative of the log density at the mode of the distribution. If this is zero, then at least locally around the mode, the log density is approximately quadratic indicating locally approximate normality of the p.d.f. Put $\eta(x;\theta) = -\frac{\partial}{\partial x}\log f(x;\theta)$. Then the above condition requires that if m satisfies $\eta(m;\theta) = 0$, it follows that $\eta''(m;\theta) = 0$. Families of distributions which are symmetric about their mode automatically satisfy the above implication but a somewhat more restrictive condition generates more interesting families of approximate normally distributed variates. For example, we may ensure the above implication by requiring that

$$\eta''(x;\theta) = p(x;\theta)\eta(x;\theta) \tag{3.31}$$

and if we take $p(x;\theta) = -\frac{c}{x^2}$ we obtain the family of probability densities of $X^{1/3}$ where X has a gamma distribution or a limit, the normal distribution.

Other interesting families result from the same differential equation with alternate choices of the function $p(x;\theta)$. The logistic family of distributions satisfy a differential equation of this form. Von Mises' circular normal distribution with probability density function of the form

$$f(x) = ce^{k\cos(x-\theta)}, \quad 0 \le x \le 2\pi, \quad 0 \le \theta \le 2\pi \tag{3.32}$$

also satisfies a differential equation of this form with $p(x;\theta)$ a constant.

Having found a family of approximately normal distributions, one can seek a normally distributed inference function by attempting to transform the observations into one of these families. In the case, for example, of a gamma (α,β) distribution, the inference function $X^{1/3} - E_{\alpha,\beta}X^{1/3}$ is close to normally distributed. However, since both the logistic distribution and the circular normal distribution above satisfy this differential equation and are significantly different in shape than the normal distribution, there is no guarantee that this simple minded technique will provide a variate close to normal. An alternative, equally simple, is to construct the family of distributions of some simple transformations such as the Box-Cox power transformations and seek a distribution satisfying the simple property that the third derivative of the log density is zero at the mode.

3.6 Background Development.

The material of this chapter covers a wide variety of topics in the theory of estimation. So only those papers of more direct relevance to our development will be mentioned here. The problem of multiple roots of the score function has been recognized for some time and treated by a number of people. Barnett (1966) is an earlier paper which is of particular interest. A more recent paper is that of Reeds (1985). The search for the global maximum of the likelihood is justifiable in the presence of Wald's conditions for global consistency but is often computationally cumbersome if the score function has large numbers of roots. The use of higher order corrections to the likelihood (c.f. Cox and Hinkley (1974), Efron and Hinkley (1978),Cox (1980), McCullagh (1984) and Skovgaard (1985)) is now being widely discussed. Much of this discussion seems to be driven by the desire to compare maximum likelihood with competitive procedures for estimation at a more delicate asymptotic level. Our analysis has been motivated by approximating E-linearity although this would appear to be only one of a number of uses to which the second order E-sufficient subspace can be put. One point to note here is that second order E-sufficiency requires that the "correction" term be a multiple of S^2-I whereas Cox and Hinkley (1974) use S^2-J. An open question here is to determine in what other ways one of these two corrections should be chosen over the other. Finally, it might be mentioned that the material in sections 3.4 and 3.5 is closely related to the construction of pivotal quantities as used for interval estimation and fiducial inference. We shall not reference that literature here. The reader is referred to Kimball (1946) and Morton (1981) for work in the context of inference function theory. However, it should also be noted that any standard normal pivotal can be regarded as an inference function.

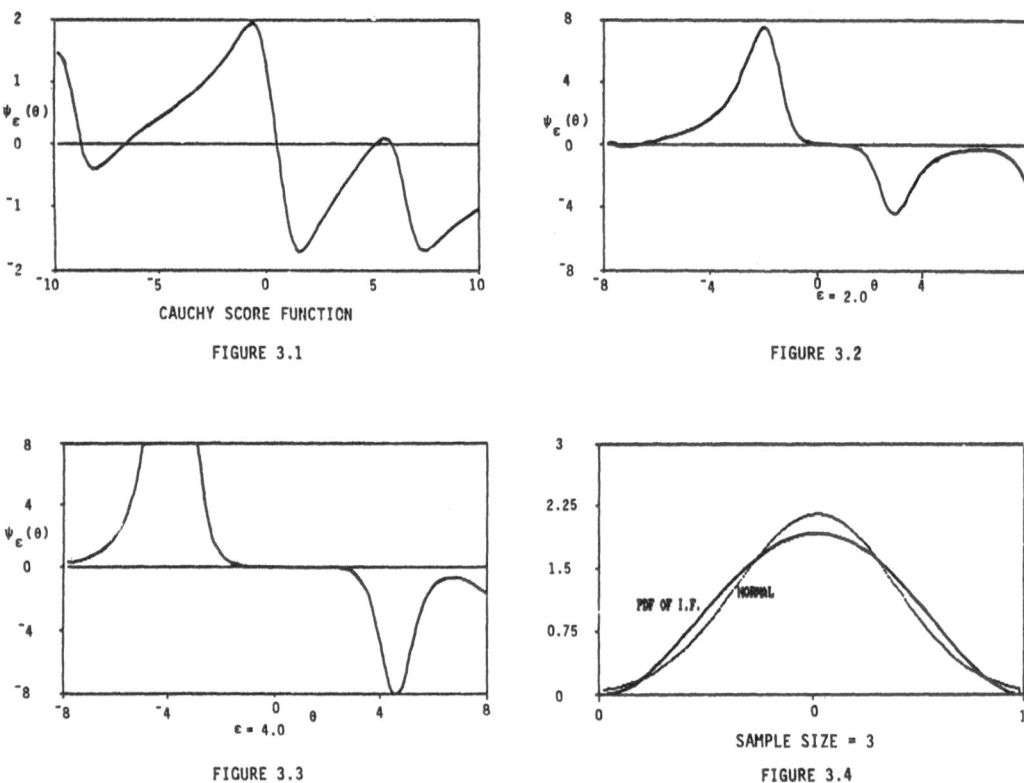

CAUCHY SCORE FUNCTION

FIGURE 3.1

FIGURE 3.2

FIGURE 3.3

FIGURE 3.4

CHAPTER 4
NUISANCE PARAMETERS

In this chapter we consider the traditional type of nuisance parameter model in which, in addition to the parameter θ we also have a vector of nuisance parameters $(\xi_1, \cdots, \xi_p)$ such that $\theta, \xi_1, \cdots, \xi_p$ is a complete description of the probability model. By this we mean that if $\theta(P) = \theta(Q)$ and $\xi_i(P) = \xi_i(Q)$ for $i = 1,...,p$ then $P = Q$. The primary problem that we shall discuss is the construction of a space Ψ with constant covariance structure for different nuisance parameter problems. In section 4.1, we consider the use of group invariant methods to eliminate nuisance parameters with particular attention to location-scale models. In section 4.2 the use of conditional models is described in the context of inference functions. Finally, in section 4.3, we shall consider the construction of inference functions in more difficult situations where techniques described earlier do not work.

4.1 Eliminating Nuisance Parameters by Invariance.

Suppose $X_1, \cdots, X_n$ come from a location-scale model with p.d.f. given by $\sigma^{-n} f[(x_1 - \theta)/\sigma, \cdots, (x_n - \theta)/\sigma]$. We treat the two parameters separately.

(a) Location Parameter. Let $a(\theta) = [\sum_{i=1}^{n}(x_i - \theta)^2]^{1/2}$ and define Ψ_1 to be the space of all unbiased square integrable functions $\psi(\theta; v)$ where

$$v = [(x_1 - \theta)/a(\theta),...,(x_n - \theta)/a(\theta)] \tag{4.1}$$

This vector will lie on an n-dimensional unit sphere with distribution dependent only on f.

(b) *Scale Parameter.* Let Ψ_2 be the space of all unbiased square integrable inference functions of the vector $w = (x_1-\bar{x}, \ldots, x_n-\bar{x})$.

Consider, for example the case where f is the standard normal product density. Then v will be uniformly distributed on the n-dimensional sphere

$$S^n = \{v=(v_1, \ldots, v_n): \sum_{i=1}^{n} v_i^2 = 1\}. \tag{4.2}$$

Define latitudes on S^n by setting

$$A_s = \{v \epsilon S: \sum_{i=1}^{n} v_i = s\} \tag{4.3}$$

for $-n^{1/2} \leq s \leq +n^{1/2}$. Then for any value η of the location parameter, the density of v on the sphere defined by centering at θ will be constant on the latitudes A_s for all s. Therefore, any function $\phi(\theta)=\phi(\theta;v)$ with the property that its average value on every latitude A_s is zero will be E-ancillary. In view of this, the set of functions of the data through $\sum_{i=1}^{n} v_i$, i.e. functions of the usual student t-statistic $t(\theta;v)=n^{1/2}(\bar{x}-\theta)/s$ is E-sufficient. Note that the expectation $E_{\eta,\sigma}t(\theta;v)$ is a linear function of η. Thus we obtain the following proposition.

4.1.1 Proposition. Suppose $X_1, X_2, \cdots, X_n$ are normally distributed with mean θ and variance σ^2. Then the student t-statistic is the unique (up to a constant multiple $k(\theta)$) E-linear function lying in the complete E-sufficient subspace of Ψ_1.

Proof. Note first that the t-statistic is complete for fixed θ over the normal family

with mean η and fixed variance σ^2. Suppose ϕ is any function in Ψ_1 for which

$E_{\eta,\sigma}\phi(\theta)=0$ for all η,σ. Then $h(t)=E_{\eta,\sigma}[\phi(\theta)|t]$ is a function of t that does not depend

upon η,σ because the distribution of v given t is parameter free. The completeness

of the t-statistic therefore implies that $h(t)=0$ a.s. The proof that $<t,\phi>_t=0$ follows

similarly to Theorem 2.7.

The E-linearity of the t-statistic is immediate. Suppose $\pi(\theta)$ is another E-linear

inference function that lies in the complete E-sufficient subspace. Then we can write

$E_\eta \pi=k_1(\theta)\dfrac{\eta-\theta}{\sigma}$ and $E_\eta t=k_2(\theta)\dfrac{\eta-\theta}{\sigma}$. Thus $k_2\pi-k_1 t$ lies in $\mathbf{S}\cap\mathbf{A}$ and must therefore be

zero. The uniqueness of the t-statistic follows.

It is interesting to note that although the t-statistic is the E-linear element up to a

multiple in the complete E-sufficient subspace, it is not the local E-sufficient function.

This can be obtained by replacing the sample standard deviation in the denominator

by the expression $a(\theta)$. It is perhaps a matter of taste as to which is preferable for

inference, although statistical tradition has definitely favored the t-statistic.

To estimate location parameters for non-normal densities $f(x)$ we can show that

functions of the form

$$\int\left[\frac{\int_0^\infty u^{n-1}f[u(x_1-\theta)+\epsilon,\ldots,u(x_n-\theta)+\epsilon]du}{\int_0^\infty u^{n-1}f[u(x_1-\theta),\ldots,u(x_n-\theta)]du}-1\right]d\Lambda_s(\epsilon) \qquad (4.4)$$

(where Λ_s has finite support and the ratios of integrals have finite second moments)

lie inside the complete E-sufficient subspace. So the locally E-sufficient function found by letting $\epsilon \rightarrow 0$ will have a root $\hat{\theta}$ satisfying

$$\int_0^\infty u^{n-1} \frac{\partial}{\partial \epsilon} f[u(x_1 - \hat{\theta}) + \epsilon, \ldots, u(x_n - \hat{\theta}) + \epsilon]|_{\epsilon=0} du = 0. \tag{4.5}$$

Consider now the case of the scale parameter of the normal model. In this case, it is easy to establish that functions in the complete E-sufficient subspace in Ψ_2 will be functions of the data through the sample variance. It can also be seen that the inference function

$$\psi(\sigma) = \frac{\sum_{i=1}^n (x_i - \bar{x})^2}{n-1} - \sigma^2 \tag{4.6}$$

is both E-linear in σ^2 and locally E-sufficient. The resulting estimator for σ^2 will therefore be the bias corrected sample variance. In the more general setting where the distribution is not assumed to be normal, we note that a locally E-sufficient function for σ can be found as the score function for σ based upon the marginal distribution of the maximal location invariant statistic.

The general group-theoretic approach can be constructed as follows. Suppose that for each value of the parameter of interest θ, the nuisance parameter model can be represented as a transformation model with group G_θ acting on the parameter space so as to leave the parameter θ invariant. Note that an alternative value θ' need not be left invariant by G_θ. Then the space Ψ for inference about θ can be constructed as the space of square integrable functions of the form $\psi(\theta, m_\theta(x))$ where $m_\theta(x)$ is the maximal invariant with respect to G_θ. For example, consider the Neyman-Scott (1948) problem in which $X_1, Y_1, \ldots, X_n, Y_n$ are $2n$ independent normal random variables with common variance σ^2 such that X_i and Y_i have common mean μ_i. We are interested in inferences about $\theta = \sigma^2$ in the presence of the nuisance parameters $\mu_1, \ldots, \mu_n$. For all θ we define $G_\theta = G$ to be the group of translations of the form

$$(x_1, y_1, \cdots, x_n, y_n) \rightarrow (x_1 + u_1, y_1 + u_1, \ldots, x_n + u_n, y_n + u_n). \tag{4.7}$$

Then the root of the locally E-sufficient function for σ^2 will be the bias-corrected maximum likelihood estimator. Although the maximum likelihood estimator is inconsistent, the root of the locally E-sufficient subspace will be consistent.

4.2 Eliminating Nuisance Parameters by Conditioning.

In this section we examine ways of eliminating nuisance parameters by conditioning on statistics. Suppose there exists a family of statistics T_θ for every $\theta \epsilon \Theta$ such that

(I) For every θ_0 and for every θ_1 the statistic T_{θ_0} is a complete statistic for the nuisance parameters $\xi_1, \cdots, \xi_p$ in the model $\{P; \theta(P) = \theta_1\}$. By this we mean that if a function h satisfies $E_P h(T_{\theta_0}) = 0$ for all P such that $\theta(P) = \theta_1$, it follows that $P[h(T_{\theta_0}) = 0] = 1$ for all such P.

We shall state results that hold for all such spaces in the presence of condition (I). Consider an inference function $\psi \epsilon \Psi$. It follows from the unbiasedness of ψ and condition (I) above that $E_\theta[\psi(\theta)|T_\theta] = 0$ almost surely. Thus ψ is conditionally unbiased, given T. It also follows immediately from condition (I) and the fact that Ψ has constant covariance structure that conditionally on T_θ any two functions ψ_1 and ψ_2 have constant covariance structure. By this we mean that $E_P[\psi_1(\theta(P))\psi_2(\theta(P))|T_\theta]$ depends upon P only through $\theta(P)$. Thus the space of functions Ψ is then seen to be a space of inference functions conditionally on the family of statistics T_θ. Furthermore, if ϕ is E-ancillary, it follows from the completeness condition (I) that $\phi(\theta;X|T_\phi)$ will be conditionally E-ancillary in the sense that $E_P[\psi(\theta;X)|T_\phi] = 0$ T- almost surely for all P, θ .

Consider now an inference function ψ which is conditionally orthogonal to a function ϕ which is, in the unconditional sense, E-ancillary. In other words,

$$E_P[\psi(\theta(P))\phi(\theta(P))|T_{\phi(P)}] = 0 \quad (a.s.\ P_\phi\ ) \tag{4.8}$$

for all P. From the remarks above, the function ϕ will be conditionally E-ancillary, given T_ϕ . Deconditioning (4.8) we observe that ψ and ϕ are unconditionally orthogonal. As ϕ is an arbitrary E-ancillary function it follows that ψ lies within the complete E-sufficient subspace of Ψ. Thus we have proved the following.

4.2.1 Proposition. If ψ is conditionally orthogonal to the set of unconditional E-ancillary functions $\mathcal{A}$, then it is also unconditionally orthogonal and hence is a member of the complete E-sufficient subspace.

The relevance of this result to the consideration of nuisance parameters is that in a number of cases by conditioning upon an appropriate statistic, a nuisance parameter can be eliminated. In such conditional models, it becomes fairly easy to construct elements of the complete E-sufficient subspace using the methods of Chapter 2. The Proposition above then shows that such a function is also an element of the complete E-sufficient subspace in the unconditional model. A condition which guarantees the elimination of nuisance parameters is the following.

(II) For every $\theta_0 \epsilon \Theta$, the statistic T_{θ_0} is sufficient for the nuisance parameters

$\xi_1, \ldots, \xi_p$ in the model $\{P : \theta(P) = \theta_0\}$.

Henceforth assume that both (I) and (II) hold. Consider all functions of the form

$$\psi(\theta) = \frac{dQ_{X \, |T_\theta}}{dP_{X \, |T_\theta}} - 1 \tag{4.9}$$

where $\theta(P) = \theta$. Note that although this appears superficially to depend on the value

of P, the sufficiency condition (II) indicates that the function is dependent only on

$\theta(P)$. We wish to show that if functions of the form (4.9) above are square integrable,

then they lie in the complete E-sufficient space of inference functions. Suppose that ϕ

is a function orthogonal to ψ given by (4.9). Then as argued before, it is conditionally

orthogonal; i.e.

$$E_P[\psi(\theta)\phi(\theta) | T_\theta] = 0 \quad a.s. \; P_\theta$$

or $E_P[\phi(\theta) | T_\theta] = E_Q[\phi(\theta) | T_\theta]$. It follows that $E_P[\phi(\theta)] = E_Q[\phi(\theta)]$ and since this is to hold

for all Q, that ϕ is E-ancillary.

There are also a few standard conditions under which, in a problem involving

nuisance parameters, a conditional score function is locally E-sufficient.

4.2.2 Proposition. If we suppose that the conditional distribution of X given T_θ

has a probability density function $f_{X | T_\theta}(x; \eta)$ satisfying the regularity conditions for

interchange of derivative and integral, then the inference function

$$\psi(\theta) = \frac{\partial}{\partial \eta} \log f_{X | T_\theta}(x; \eta) \big|_{\eta = \theta} \tag{4.10}$$

is locally E-sufficient.

Proof. Suppose $\phi(\theta)$ is an unbiased inference function satisfying the condition

$E_P[\phi(\theta)\psi(\theta)] = 0$ for all P, $\theta = \theta(P)$. Then arguing again from the completeness of $T_\bullet$,

$E_\bullet[\phi(\theta)\psi(\theta)|T_\bullet] = 0$ $T_\bullet$-almost surely. Therefore in the conditional model given T, the

local E-sufficiency of the score function in that model ensures that

$\frac{\partial}{\partial \eta} E_\eta[\phi(\theta)|T_\bullet]\,|_{\eta=\theta} = 0$, $T_\bullet$-almost surely. But under conditions allowing the interchange

of integral and derivative below, suppose P_η is a set of probability measures satisfying

$\theta(P_\eta) = \eta$. Then

$$\frac{\partial}{\partial \eta} E_\eta[\phi(\theta)]\,|_{\eta=\theta} = \frac{\partial}{\partial \eta} \int E_\eta[\phi(\theta)|T_\bullet]dP_\eta\,|_{\eta=\theta} \tag{4.11}$$

$$= \int \{\frac{\partial}{\partial \eta} E_\eta[\phi(\theta)|T_\bullet]\}dP_\eta\,|_{\eta=\theta} + E_\bullet\{E_\bullet[\phi(\theta)|T_\bullet]\frac{\partial}{\partial \eta}\frac{dP_\eta}{dP_\bullet}\,|_{\eta=\theta}\}.$$

Now the first term in the last expression above is zero since $\frac{\partial}{\partial \eta} E_\eta[\phi(\theta)|T_\bullet]\,|_{\eta=\theta} = 0$ and

the second term is zero since $E_\bullet[\phi(\theta)|T_\bullet] = 0$. Therefore, the function ϕ is locally ancil-

lary; it follows that the function ψ is locally E-sufficient.

4.3 Inference For Models Involving Obstructing Nuisance Parameters.

We have coined the term obstructing nuisance parameter for those nuisance

parameters in which there is no obvious way to construct an inference function space

of unbiased functions possessing constant covariance for the purposes of inference as

outlined in chapter 2. Typically such problems arise from the presence of nuisance

parameters whose effect on first and second moments of functions cannot be elim-

inated. We shall leave the main recommendation for handling such problems to later,

and shall first consider an argument for relaxing the unbiasedness criterion for inference functions as a mechanism for solving the problems of obstructing nuisance parameters. Our major recommendation will be less drastic than the elimination of unbiasedness, but for the sake of completeness it is appropriate to consider whether inference functions which are not unbiased have a role in inference. For example, consider a set of n independent normal variates with mean μ and unit variance. Define a parameter of interest θ by setting $\theta = 0$ for $\mu < 0$, and $\theta = 1$ for $\mu \geq 0$. The absolute value of μ is regarded as a nuisance parameter. Now suppose that $\psi(\theta; X_1, \cdots, X_n)$ is an unbiased inference function for θ. Then we observe that $E_\mu[\psi(\theta(\mu)) | \bar{X}] = 0$ almost surely for all μ because $\bar{X}$ is a complete statistic for the nuisance parameter. However, $\bar{X}$ is also sufficient for μ, and therefore ψ must be E-ancillary. Thus there would seem to be no practical way to make inferences about θ through the use of unbiased inference functions. A sensible inference function in this context would be $\psi(\theta) = 1_{(\bar{X} \geq 0)} - \theta$. Unfortunately, this function is not unbiased. Nevertheless, there is a generalization of the concept of unbiasedness which admits this function. Consider for the moment a general model with parameter of interest θ and nuisance parameters $\xi = (\xi_1, \ldots, \xi_p)$. Let $\psi(\theta; X)$ be given but not necessarily be unbiased. We now suggest the following generalization.

4.3.1 Definition. A function ψ is said to be *bias consistent* if for every value of θ, η and ξ,

$$| E_{\eta, \xi} \psi(\theta) | \geq | E_{\theta, \xi} \psi(\theta) |. \tag{4.12}$$

4.3.2 Definition. A function ψ is said to be *bias ancillary* if for every value of θ, η and ξ,

$$E_{\eta,\xi}\psi(\theta = E_{\theta,\xi}\psi(\theta) .$$
(4.13)

In the normal example given above, we see that every unbiased function is E-ancillary, and therefore bias ancillary. However, the function $\psi(\theta) = 1_{(\bar{x} \geq 0)} - \theta$ is bias consistent but not bias ancillary.

Within the context of nuisance parameter problems, the example given above must be considered fairly extreme. We now examine a set of local techniques for inference which represent a more modest compromise of the basic structure of the inference function space. As we have seen, the construction of unbiased functions for models involving nuisance parameters is closely related to the construction of unbiased estimators of zero. This cannot be accomplished non-trivially in all settings. The difficulty would, of course, be removed if we were to write our inference functions as functions of both the parameter of interest and the nuisance parameter: $\psi(\theta,\xi)$. For example the components of the multiparameter score vector are of this form. Unbiasedness then requires that $E_{\theta,\xi}\psi(\theta,\xi) = 0$ for all values of the parameters. To estimate θ in this case, we can construct an estimator $\tilde{\xi}(\theta)$ of ξ and then solve the equation $\psi(\hat{\theta},\tilde{\xi}(\hat{\theta})) = 0$. Aside from the fact that the estimator $\hat{\theta}$ so obtained will depend on the rather arbitrary choice of an estimator of ξ, there is the additional difficulty posed by the fact that $E_{\theta,\xi}\psi(\theta,\tilde{\xi}(\theta))$ need not be zero. It is this that gives rise to the Neyman-Scott paradox involving joint maximization over parameters.

The problem seems to arise in part because the expectation of the function ψ is sensitive to perturbations in the nuisance parameter.

This suggests that we may permit functions of both parameters but require that $E_{\theta,\eta}\psi(\theta,\xi) = 0$ for all choices of θ,η and ξ. Of course, this brings us full circle to the original problem of constructing unbiased functions of θ. However, in this context, where the functions depend upon the nuisance parameter, we can relax the unbiasedness in the following way. Consider for the moment the case where the nuisance parameter ξ is one-dimensional.

4.3.3 Definition. The function $\psi(\theta,\xi)$ is said to be *locally unbiased* if $E_{\theta,\xi}\psi(\theta,\xi){=}0$ for all θ and ξ, and if it is locally E-ancillary for ξ for each value of θ, or more formally,

$$\frac{\partial}{\partial\eta}E_{\theta,\eta}\psi(\theta,\xi)\mid_{\eta=\xi} = 0. \tag{4.14}$$

The approach of Chapter 2, establishing a space of constant covariant structure, is preferable when the space is sufficiently rich to allow satisfactory inference, because, in a sense, this space eliminates the nuisance parameter. On the other hand, when this space cannot be found or is not sufficiently rich, an alternative approach through the above mentioned locally unbiased functions may be tried. An inference function $\psi(\theta,\xi)$ is locally unbiased if it is orthogonal to S_2 at each value of the parameter, where $S_i(\eta_1,\eta_2) = \frac{\partial}{\partial\eta_i}\log f(x;\eta_1,\eta_2)$. It is not difficult to show that the function

$$S_1(\theta,\xi) - c(\theta,\xi)S_2(\theta,\xi) \tag{4.15}$$

where

$$c(\theta,\xi) = \frac{Cov(S_1,S_2)}{Var(S_2)} \tag{4.16}$$

is locally unbiased.

Similarly, a function is *second order unbiased* if it is orthogonal to both S_2 and $S_2^2 - I_2$, where

$$I_2 = -\frac{\partial}{\partial \xi} S_2(\theta,\xi) \tag{4.17}$$

is the observed information function. One such function is

$$S_1 - c_1 S_2 - c_2(S_2^2 - I_2) \tag{4.18}$$

where

$$\underline{c} = (c_1,c_2)' = \Sigma^{-1} B \tag{4.19}$$

where Σ is the covariance matrix of $(S_2, S_2^2 - I_2)$ and B is the vector of covariances between S_1 and $(S_2, S_2^2 - I_2)$.

4.3.4 Example. As an illustration of the construction of locally unbiased functions, consider the estimation of the mean and variance in the normal model using components of the score function. The inference function for μ, namely

$$\sum_{i=1}^{n} (x_i - \mu)/\sigma^2 \tag{4.20}$$

is seen to be unbiased. However, the function for σ

$$\frac{-n}{\sigma} + \frac{\sum_{i=1}^{n}(x_i - \mu)^2}{\sigma^3} \tag{4.21}$$

is locally (first order) unbiased but not second order unbiased. Introducing the correction for second order bias given above, we obtain the bias corrected inference function.

$$\frac{-(n-1)}{\sigma} + \frac{\sum\limits_{i=1}^{n} (x_i - \bar{x})^2}{\sigma^3}. \tag{4.22}$$

We have already seen that in order to utilize the locally unbiased functions available for inference we have to construct an estimate for the nuisance parameter. So unlike the framework developed in Chapter 2, in which nuisance parameters were eliminated up to their second moment properties (i.e., constant covariance) we must now consider them in order to estimate the parameter of interest. Thus it can be seen that what are needed here are techniques of inference that allow joint estimation of all the parameters of the model with the qualification that each parameter in turn is to be regarded as a parameter of interest to be separated out from the remaining "nuisance" parameters. Another way of saying this is that we wish to estimate all parameters jointly in such a way as to discount in the estimation of any given parameter the effects of estimating the other parameters. Under such circumstances, inference functions are most appropriately understood as vector valued functions whose ranges have the same dimension as the dimension of the parameter. Suppose $\mathcal{P}$ is a k-parameter model in the sense that $\theta = (\theta_1, \cdots, \theta_k)$ totally determines the underlying distribution of the data X. We will call a function $g(\eta_1, \eta_2, \ldots \eta_r)$ *locally constant to order m* at a point θ if all (mixed) partial derivatives of the function of order less than or equal to m are zero when evaluated at the point θ.

4.3.5 Definition. Consider functions for inference of the form $\psi(\theta;X)=(\psi_1(\theta;X), \ldots, \psi_k(\theta;X))$ where for all θ we impose the *unbiasedness condition* that $E_\theta\psi(\theta;X)=0$ and the condition that for every i $\psi_i(\theta;X)$ is *m−th order locally unbiased* for θ_i. By this we mean that $E_{\eta_1\eta_2\cdots\eta_k}\psi_i(\theta;X)$ is locally constant to order m jointly in all variables $\eta_j, j\neq i$ at the point where $\eta_1=\theta_1,\ldots,\eta_k=\theta_k$. We shall call such a space a *m−th order local inference space*.

4.3.6 Definition. A function ϕ within this space will be said to be *m−th order locally N-ancillary* (N for "nuisance") if for all i $E_{\eta_1,\ldots,\eta_k}\phi_i(\theta;X)$ is *locally constant to order m jointly in all k variables* $\eta_1,\ldots,\eta_k$ at the point where $\eta_1=\theta_1,\ldots,\eta_k=\theta_k$.

A function ψ in the *m−th* order local inference space will be said to lie in the *m−th order locally N-sufficient subspace* if

$$E_{\theta_1,\ldots,\theta_k}\psi_i(\theta;X)\phi_i(\theta;X)=0 \tag{4.23}$$

for every $i=1,\ldots,k$, and every *m−th* order locally N-ancillary function ϕ.

We are now in a position to examine the consequences of inference of various orders within these local inference spaces. We consider the first order.

4.3.7 Proposition. Consider the first order inference space for a model with parameters θ_1 and θ_2. We assume the usual regularity that allows the interchange of derivatives and integrals. Let $c_{ij}=\dfrac{Cov(S_i,S_j)}{Var(S_j)}$. Then the inference function

$$\psi(\theta_1,\theta_2;X) = (S_1-c_{12}S_2,S_2-c_{21}S_1) \tag{4.24}$$

generates the first order N-sufficient subspace.

Proof. As noted in equation (4.15), ψ is in the first order local inference space. Suppose ϕ is such that for $i=1,2$ we have $E_{\theta}\psi_i(\theta)\phi_i(\theta)=0$. Then for $i\neq j$ we obtain $E_{\theta}S_i\phi_i=c_{ij}E_{\theta}S_j\phi_i$. This identity can be rewritten as

$$\frac{\partial}{\partial\eta_i}E_{p_1,\eta_2}\phi_i(\theta_1,\theta_2)|_{\eta_1=\theta_1,\eta_2=\theta_2} = c_{ij}\frac{\partial}{\partial\eta_j}E_{\eta_1,\eta_2}\phi_i(\theta_1,\theta_2)|_{\eta_1=\theta_1,\eta_2=\theta_2}. \tag{4.25}$$

But the right hand side of this identity is zero because ϕ is locally unbiased. Therefore it follows that ϕ is first order N-ancillary. The converse, namely that the first order N-ancillarity of ϕ implies that it is uncorrelated with ψ, follows similarly.

It can be seen that the equations $\psi=0$ do not lead to a new pair of estimators for θ_1 and θ_2 because they are solved by $S_1,S_2=0$, which yields the maximum likelihood estimators for θ_1 and θ_2. This does not carry over to the higher order examples. In these cases, estimators other than the maximum likelihood estimators will result. We can write an element of the second order N-sufficient subspace as $\psi=(\psi_1,\psi_2)$, where

$$\psi_i = \sum_{j=1}^{2} a_{ij}S_j + \sum_{j=1}^{2}\sum_{k=1}^{j} b_{ijk}(S_jS_k-I_{jk}). \tag{4.26}$$

There are a total of 10 coefficients in (4.26), but there are also four constraints that must be imposed to ensure that ψ is second order unbiased. The functions satisfying these constraints generate the second order N-sufficient subspace of the second order local inference space. Within this space one might choose a function such as a locally E-linear function analogously to section 2 of chapter 3. The total set of 10 equations that the coefficients must be chosen to satisfy in this case will be for $i,j,k=1,2,$

$$E_{\iota_1 \iota_2} \frac{\partial^2}{\partial \theta_i^2} \psi_i(\theta_1, \theta_2) = 0 \tag{4.27}$$

$$\frac{\partial}{\partial \eta_j} E_{\eta_1 \eta_2} \psi_i(\theta_1, \theta_2)|_{\eta_1 = \iota_1, \eta_2 = \iota_2} = 0, \quad j \neq i.$$

$$\frac{\partial^2}{\partial \eta_i \partial \eta_j} E_{\eta_1 \eta_2} \psi_k(\theta_1, \theta_2)|_{\eta_1 = \iota_1, \eta_2 = \iota_2} = 0, \; i \leq j.$$

The first pair of equations in (4.27) is analogous to condition (I) of section 3.2. The second pair of equations induces second order local unbiasedness, and the third family of six equations is analogous to condition (III) of section 3.2.

4.3.8 Example. Suppose that n independent normal variates with mean μ and variance σ^2 are observed. Let $\theta_1 = \mu$ and $\theta_2 = \sigma^2$. Then the second order local inference function lying in the second order N-sufficient subspace satisfying the linearity conditions of (4.27) is

$$\psi = (\psi_1, \psi_2) = [\sum_{i=1}^{n} (x_i - \mu), \frac{\sum_{i=1}^{n} (x_i - \bar{x})^2}{n-1} - \sigma^2]. \tag{4.28}$$

This function can easily be checked to satisfy the required conditions.

4.4 Background Details.

Godambe and Thompson (1974) and Godambe (1976) examined the use of inference functions in nuisance parameter settings, and showed that the theory had the potential to remove the awkwardness of maximum likelihood estimation that appears in its most extreme form as inconsistency in the Neyman-Scott problem. Other papers include Chandrasekar and Kale (1984), Ferreira (1982a,b), Godambe (1984) and Okuma (1975,1977). Of related interest is the extension of the theory of inference functions or estimating equations to the multiparameter setting. Work in this area includes Ferreira (1982), Kale (1962) and numerous applications, of which McLeish (1984) is one.

CHAPTER 5
INFERENCE UNDER RESTRICTIONS:
LEAST SQUARES, CENSORING AND ERRORS IN VARIABLES TECHNIQUES

5.1 Linear Models.

There are many problems in which a solution would be relatively easier to obtain if there were some additional information recorded or observable. We may, for example, observe a sum of variates of the form $Y = X + \epsilon$ where X and ϵ are independent. In general, if our observations are distributed according to a convolution, the probability density function may be intractable for maximum likelihood estimation since it may be expressible only as an integral or sum. However, if the components of the sum were observable, then estimation by likelihood methods would often be quite easy.

Censoring is another common mechanism whereby problems, easily solvable in its absence, become substantially more difficult to handle by likelihood methods. In fact, when the censoring mechanism has a stochastic component, the full likelihood would require a stochastic model for this mechanism, and this is clearly undesirable when the parameters of interest are those of the failure distribution, not the censoring distribution. Such a problem is often solved by dealing with a *partial* or *marginal* likelihood and/or by making assumptions on the interaction between the censoring and the failure variables.

Our approach to such problems, consistent with that of previous sections, is to limit the space of potential inference functions under consideration to a space more tractable and then carry out reductions by E-sufficiency or local E-sufficiency. Once again, this is done in part to keep the regularity, often required in the solution of a problem, imposed on the space of inference functions over which we exert control rather than on the mechanism underlying the generation of the data which is clearly beyond our control. The next few examples are models in which *conditional least squares* may be applied:

5.1.1 Example. Consider square integrable random variables X_i, $i=1, 2, \ldots n$ adapted to an increasing family of σ-fields, F_i. Suppose the conditional expectation of X_i is a linear function of the parameter θ. In other words, for a F_{i-1} measurable random variate μ_i whose expectation depends only on the parameter θ,

$$E_\theta[X_i \,|F_{i-1}] = \mu_i\theta. \tag{5.1}$$

Let Ψ be the space of inference functions of the form

$$\psi(\theta)= \sum_{i=1}^{n} a_i(\theta)(X_i-\mu_i\theta) \tag{5.2}$$

where $a_i(\theta)$ is a non-random weight function. Suppose the conditional variance

$$Var_P[X_i \,|\, F_{i-1}]= \Sigma_i$$

has positive (finite) expectation and depends only on P through $\theta(P)$. The nuisance parameter here is P, the actual joint distribution of the observations. However, the conditions insure that the covariance between two members of Ψ is a function of P only through $\theta(P)$ and so the space Ψ has constant covariant structure. Then the function

$$\psi^*(\theta)= \sum_{i=1}^{n} \frac{E_\theta\mu_i}{E_\theta\Sigma_i}(X_i - \mu_i\theta) \tag{5.3}$$

is locally E-sufficient in this space.

Proof. Denote the coefficients of the locally E-sufficient function above by $a_i^{\bullet}(\theta)$. Suppose a function ψ given by (5.2) is orthogonal to this function. It follows that

$$\sum_{i=1}^{n} a_i a_i^{\bullet} E_{\theta} \Sigma_i = 0.$$

Therefore,

$$\sum_{i=1}^{n} a_i(\theta) E_{\theta} \mu_i = \frac{d}{d\eta} E_{\eta} \psi(\theta)\big|_{\eta=\theta} = 0$$

and therefore the function ψ is locally E-ancillary.

5.1.2 Example. Consider square integrable random variables $X_1, X_2, \cdots$ from a one parameter family with parameter θ. Suppose X_i is adapted to an increasing family of σ-fields, F_i and the conditional expectation of X_i is a linear function of the unknown parameter θ. In other words, for some set of F_{i-1} measurable random variates μ_i,

$$E_{\theta}[X_i \mid F_{i-1}] = \mu_i \theta. \tag{5.4}$$

Let Ψ be the space of inference functions of the form

$$\psi(\theta) = \sum_{i=1}^{n} A_i(\theta)(X_i - \mu_i \theta) \tag{5.5}$$

where $A_i(\theta)$ are F_{i-1} measurable random variables such that the function $\psi(\theta)$ is square integrable. Suppose we define the conditional variance

$$Var_{\theta}[X_i \mid F_{i-1}] = \Sigma_i(\theta)$$

Then if the function

$$\psi^{*}(\theta)= \sum_{i=1}^{n}\mu_{i}\Sigma_{i}^{-1}(\theta)(X_{i} - \mu_{i}\theta) \tag{5.6}$$

is in the space Ψ , it generates the locally E-sufficient subspace.

Proof If $\psi(\theta)$ is of the form (5.5) and ψ^* is given by (5.6), we need to show

$E_{\theta}\{\psi^*(\theta)\psi(\theta)\} = 0$ implies $\frac{d}{d\xi}\{E_{\xi}\psi(\theta)\}|_{\xi=\theta} = 0$. Observe that

$$E_{\theta}\{\psi^*(\theta)\psi(\theta)\} = E_{\theta}\sum_{i=1}^{n}\mu_{i}\Sigma_{i}^{-1}(\theta)\Sigma_{i}(\theta)A_{i}(\theta)$$

$$= E_{\theta}\sum_{i=1}^{n}\mu_{i}A_{i}(\theta)$$

$$= \frac{d}{d\xi}E_{\xi}\sum_{i=1}^{n}(\xi-\theta)\mu_{i}A_{i}(\theta)|_{\xi=\theta}$$

$$= \frac{d}{d\xi}E_{\xi}\psi(\theta)|_{\xi=\theta}$$

This example above is closely related to the theory of quasi-likelihood estimation which will be discussed further in Chapter 6.

5.1.3 Example. Consider the model of example 5.1.2 only now we seek an E-sufficient (rather than locally E-sufficient) subspace. In particular, suppose $\Sigma_{i} = \Sigma_{i}(\theta)$ does not depend on θ . We limit ourselves to the subspace of functions of the form (5.5) with A_{i} also independent of θ . Then in this space Ψ of functions,

$$\psi^{*}(\theta) = \sum_{i=1}^{n}\mu_{i}\Sigma_{i}^{-1}(X_{i}-\mu_{i}\theta) \tag{5.7}$$

generates the complete E-sufficient subspace.

5.1.4 Example Suppose we consider n unbiased inference functions

$\psi_1(\theta)$, $\psi_2(\theta)$, ... $\psi_n(\theta)$ with non-singular covariance matrix and let Ψ be the space of linear combinations of the form

$$\sum_{i=1}^{n} a_i(\theta)\psi_i(\theta) \tag{5.8}$$

with arbitrary non-random coefficients $a_i(\theta)$. Suppose that

$$Cov_P\left(\psi_i(\theta),\psi_j(\theta)\right) \tag{5.9}$$

is constant for all P such that $\theta(P) = \theta$. This implies that the space has constant covariant structure in the nuisance parameter. Then the inference function

$$\psi_i^*(\theta) = \sum_{i=1}^{n} a_i^*(\theta)\psi_i(\theta)$$

is locally E-sufficient in this space where

$$a^*(\theta) = \Sigma^{-1}(\theta)\frac{\partial}{\partial \eta}E_\eta\psi(\theta)|_{\eta=\theta} \tag{5.10}$$

$$\Sigma_{ij}(\theta) = cov_\theta(\psi_i(t),\, \psi_j(t))$$

and

$$\psi(\theta)=(\psi_1(\theta),\, \ldots,\psi_n(\theta))^T.$$

In particular, suppose the random sample X_i consists of independent identically distributed random variables with cumulative distribution function $F_\theta(x)$. Then if $\Phi(x)$ denotes the standard normal cumulative distribution function, there is an essentially unique class of inference functions (5.8) such that it has, for each fixed θ, a normal distribution and it is monotone in each component X_i. These functions can be written in the form

$$\sum_{i=1}^{n} a_i(\theta)\Phi^{-1}(F_\theta(X_i)).$$

Within this class of functions, the function

$$\sum_{i=1}^{n}\Phi^{-1}(F_\theta(X_i))$$

generates the complete E-sufficient subspace. This case illustrates the fact that spaces of functions can be built which have the useful feature that there elements are pivotal quantities. In view of the fact that linear combinations of random variables are being used, the normal assumption above is particularly natural. Such a technique can also be used on data sets in which the random variables do not have finite moments. The transformation to normality not only ensures the existence of moments but makes precise pivotal methods possible.

Another prominent example is that of *L-estimators*. In this case, we consider a location family with unknown location parameter θ and known covariance structure among the order statistics. Here, $\psi_i(\theta) = X_{(i)} - \theta$ and the optimal coefficients can be approximated in the case of known probability density function $f(x-\theta)$ by

$$a_i^*(\theta) \approx \{\frac{f''}{f} - (\frac{f'}{f})^2\}(F^{-1}(\frac{i}{n-1})),$$ where F is the cumulative distribution function of f .

For another example , suppose we observe independent X_i from a location family with a symmetric probability density function $f(x-\theta)$ known up to the location parameter. Let Ψ be the space of *linear rank statistics* of the form

$$\frac{1}{n}\sum_{i=1}^{n}a(R_i) \tag{5.11}$$

where $R_i = R_i(\theta)$ is the rank of X_i among the values $X_1,X_2,.....,X_n,2\theta-X_1,2\theta-X_2,\ldots,2\theta-X_n$

and $a(R_i)$ is an arbitrary score function such that $\sum_{i=1}^{2n} a(i) = 0$. Then the locally E-sufficient rank statistic is the function of the form (5.11) with

$$a(R_i) = E_0[S(X_i)|R_i]$$

where $S(x) = \dfrac{-f'(x)}{f(x)}$. Once again this can be approximated locally by

$$a(R_i) \approx S(F_0^{-1}(\frac{R_i}{2n}))$$

with F_0^{-1} the inverse of the cumulative distribution function of the variates X_i under the parameter $\theta = 0$.

5.2 Censoring, Grouping and Truncation.

The observation of incomplete data is perhaps more the rule than the exception in statistical inference. Theoretical considerations may make the assumption of a given model appealing for the actual data but the limitations of the observation process mean that we observe only some function of the data, perhaps including random noise. This is particularly true in any model in which we assume a given relation between variates such as a time series or regression model. The dependence assumed here clearly depends on the actual values not on their observed values if error is introduced to the observations.

Suppose, for example, the actual data X has some assumed distribution described by a probability density function $f_\theta(x)$ but we are unable to observe X precisely, and instead see only some transformation of X, say T. Typically, T may be X rounded or truncated, and often the extent of the rounding will depend on the value of X itself, so

the transformation will be by no means a simple one. Many such transformations, however, can be written as a *linear* operator on the space of random variables and the
most common such operator is the conditional expectation. For example if the rounding
partitions the space into a sigma field G, we will observe the rounded random variate
$T = E[X|G]$. The question we consider in this section is the effect that such a reduction should have on the appropriate inference function and the estimator defined
thereby.

We begin with an example of a projection argument applied to a symmetrized
inference function discussed in chapter 3.

5.2.1 Example .

We have seen in the general one-parameter problem that inference functions of
the form

$$\psi_\xi(\theta) = \frac{L(\xi)}{L(\theta)} - 1$$

span the space of E sufficient inference functions. Since we are concentrating on the L_2
space of square integrable functions with its associated metric, it seems natural also to
attempt to provide inference functions that are approximately normally distributed,
since for normally distributed variates, maximum likelihood is carried out by minimiz-
ing the L_2 norm. Demanding that the inference functions are normally distributed by
imposing conditions on their cumulants is one possibility here. A simpler alternative
is to attempt to symmetrize the distribution of $\psi(\theta)$ hoping that in so doing, we obtain
an approximately normal pivotal. One of the simplest criteria for symmetrization of a

distribution is to require that its median and mean are equal.

Suppose, for example, $X_1, X_2, ..., X_n$ are independent identically distributed random variates with an exponential distribution, mean θ. The inference functions generating the E sufficient subspace are

$$\psi_\xi(\theta) = (\theta/\xi)^n \exp\{-S_n[\xi^{-1}-\theta^{-1}]\}-1 \tag{5.12}$$

where $S_n = \sum_{i=1}^{n} X_i$. If we choose ξ in such a way that the median of this variate is 0 under the parameter θ, it is easy to see that $\xi=\theta\xi_1$ where ξ_1 satisfies the equation

$$n \log\xi_1 = m_n[1-\xi_1^{-1}]$$

and where m_n is the median of a gamma distribution with shape parameter n and scale parameter 1. Good approximations are available for the median of a Gamma variate and so ξ_1 can be easily approximated. One simple formula that gives a reasonable approximation is $\xi_1 = 2^{-1/n}$. With this value of ξ_1, the inference function ψ_ξ is given by

$$\psi(\theta) = \xi_1^{n[\frac{S_n}{\theta m_n}-1]} - 1 \approx 2^{1-\frac{S_n}{\theta m_n}}-1. \tag{5.13}$$

Now let us suppose that c of the above n observations are right censored according to type I censorship. In particular, we observe n-c of the values X_i , those for which $X_i \leq L_i$ where L_i is an associated sequence of independent censoring times. Since $\psi(\theta)$ is now approximately normally distributed (see graphs of probability density functions in figures X) it seems reasonable to condition the above inference function on the observed information $\{c, \min(X_i,L_i); i=1,2,...,n\}$. With $T = \sum_{i=1}^{n} \min(X_i,L_i)$, the resulting inference function is

$$\psi^*(\theta) = \xi_1^{c-n} \exp{-T/\theta} \; [\xi_1^{-1} - 1] - 1 = \xi_1^{\frac{nT}{\theta m_n} - n + c} - 1. \qquad (5.14)$$

Standardizing this to obtain an approximately N(0,1) pivotal, we obtain the function

$$\frac{\psi^*(\theta)}{[\xi_1^{c-n}(2 - \xi_1)^{c-n} - 1]^{1/2}} \qquad (5.15)$$

This pivotal may be compared with that of Sprott (1973) for the censored exponential distribution

$$\frac{\hat{\theta}^{-1/3} - \theta^{-1/3}}{[\hat{\theta}^{-2/3}/9(n-c)]^{1/2}} \qquad (5.16)$$

where $\hat{\theta} = T/(n-c)$.

Separating the questions of normality and standardization in the pivotal (5.16), the normality of this expression is equivalent to the normality of the pivotal $(\hat{\theta}/\theta)^{1/3}$ whose probability density function is given by

$$f(x) = \frac{3r^r}{(r-1)!} x^{3r-1} e^{-rx^3}, \quad x > 0. \qquad (5.17)$$

where $r = n-c$.

On the other hand the normality of the pivotal ψ^* is equivalent to that of e^{-Y} where Y has a Gamma distribution with parameters r and $\xi_1^{-1} - 1$.

While our objective of symmetrization is fulfilled (the distribution of e^{-Y} is very close to symmetric about its mode for all values of $r \geq 1$), the transformation $T^{1/3}$ produces a more normal looking probability density function for virtually all values of $r \geq 1$. A more direct approach to achieving normality of the inference function is to use the exactly normal pivotal

$$\Phi^{-1}(F_r(\tfrac{T}{\theta}))$$

where F_r and Φ are the gamma $(r,1)$ and standard normal cumulative distribution functions respectively. This pivotal is very close to the standardized variate $T^{1/3}$. There appear to be, however, few such models where a single transformation independent of a shape parameter provides near normality over the whole range of values of a shape parameter.

5.3 Errors in Observations.

Consider first the case where $Y = X + \epsilon$ where ϵ is a random variable independent of X. We suppose that an unbiased inference function of X is available, say $\psi(\theta, X)$ (note that we are now showing both arguments of the inference function explicitly) and we seek a function appropriate for Y. Expanding by a Taylor series expansion to the second order,

$$\psi(\theta, X) \approx \psi(\theta, Y) - \epsilon \frac{\partial}{\partial x} \psi(\theta, Y) + \tfrac{1}{2} \epsilon^2 \frac{\partial^2}{\partial x^2} \psi(\theta, Y)$$

Taking conditional expectations with respect to Y,

$$E_\theta[\psi(\theta, X) \mid Y] \approx \psi(\theta, Y) - E_\theta[\epsilon \mid Y] \frac{\partial}{\partial x} \psi(\theta, Y) + \tfrac{1}{2} E_\theta[\epsilon^2 \mid Y] \frac{\partial^2}{\partial x^2} \psi(\theta, Y). \tag{5.18}$$

Since the distribution of ϵ is generally unknown, it seems reasonable to estimate $E_\theta[\epsilon^k \mid Y]$ using a regression on the powers of Y. To first order, these conditional expectations may be replaced by

$$E_\theta[\epsilon \mid Y] \approx E(\epsilon) + b(\theta)(Y - E_\theta Y) \tag{5.19}$$

where

$$b(\theta) = \frac{Var_\theta(\epsilon)}{Var_\theta(Y)}$$

and

$$E_\theta[\epsilon^2|Y] \approx E_\theta^2[\epsilon|Y] + (1-b)Var(\epsilon) \tag{5.20}$$

Substituting these expressions in (5.18) gives an approximately unbiased inference function of Y.

5.3.1 Example. Let X have a normal $(\theta,1)$ distribution and ϵ have an arbitrary distribution with known expected value $E(\epsilon) = \mu$ and variance σ^2. Suppose ψ is the score function for X, $\psi(\theta,X) = X - \theta$. Then from (5.18) with the approximation (5.19), the adjusted inference function is

$$\frac{Y-\theta-\mu}{\sigma^2+1} \tag{5.21}$$

which, interestingly, is the score function computed from the marginal distribution of Y *assuming that the errors (and hence Y) are normally distributed.*

5.3.2. Example. Suppose X and ϵ are distributed independently, X distributed $N(0,\theta^2)$ and ϵ has an arbitrary distribution with known mean 0 and variance σ^2. Let ψ be the score function when we observe X exactly, i.e. $\psi(\theta,X) = \frac{X^2-\theta^2}{\theta^3}$. Substituting the approximations (5.19) and (5.21) in the expansion (5.18), we obtain after some simplification

$$\frac{\theta}{(\theta^2+\sigma^2)^2}[Y^2-\theta^2-\sigma^2] \tag{5.22}$$

Clearly, the resulting inference function is unbiased and asymptotic to $\psi(\theta,X)$ as

$\sigma^2 \to 0$.

In the above cases, we assumed that we knew the moments (first and second) of

the error distribution, and in practice they may well be unknown. Fortunately, there is

an alternative approach to handling these errors hinted at in chapter 4 through con-

cepts of local unbiasedness. In the above examples, we arrived at an inference function

that was a linear combination of the functions ψ, ψ', ψ'', $Y\psi'$, $Y^2\psi''$ all normalized to have

zero expectation where the indicated derivatives are with respect to the second com-

ponent of the function (e.g. $\psi'' = \dfrac{\partial^2}{\partial x^2}\psi(\theta,Y)$. A more direct approach to the problem

might have been to seek coefficients of these terms $c_i(\theta)$ such that the resulting func-

tion

$$\psi + c_1\psi' + c_2Y\psi' + c_3\psi + c_4Y\psi'' + c_5Y^2\psi'' + c_6$$

is locally unbiased, treating the moments of the error μ, σ^2 as nuisance parameters, the

analysis conducted locally around $\mu = 0$ and $\sigma^2 = 0$, with the shape of the error distri-

bution known. In this case, since we have five free coefficients to select from, we

might expect to be able to find coefficients such that the function is fifth order locally

unbiased in a general one nuisance parameter case and this should insure a high degree

of insensitivity to local changes in the value of the nuisance parameter.

5.4 Background Details.

The foundations of the theory of conditional least squares for stochastic processes was developed by Klimko and Nelson (1978). The use of conditional least squares for estimation in aggregate data has been suggested by Kalbfleisch and Lawless (1984). McLeish (1983,1984) has proposed a systematic approach to these methods through the use of inference function tools. The theory of this chapter can be considered an elaboration of these methods. A recent paper in a similar vein is Firth (1987). Godambe (1985) proposed the use of his optimality criterion for inference on stochastic processes. While the term " conditional least squares" may be a misnomer in this context, it has become popular. The use of such methods provides criteria for estimation that are computationally more tractable than maximum likelihood and avoid some of the problems associated with nuisance parameters.

Likelihood methodology is justifiable on a discrete probability space line by a simple intuitive interpretation of likelihoods. For example, in a discrete probability space, the maximum likelihood estimator is that value of the parameter that would render the probability of the data maximum. However, when our observations are realized in a more complicated space, the intuitive argument for likelihood methodology is more difficult. In the case of models for continuous time stochastic processes there is often no single dominating measure on a sample space of paths that is analogous to Lebesgue measure for continuous random variables. Thus maximum likelihood estimates cannot be generally computed by maximizing over a convenient likelihood function. This leads to difficulties both in computation and in stability of the procedure as we shall now discuss. Consider for example a Wiener process X_t with zero drift and diffusion parameter σ. Suppose we wish to estimate σ based upon a single realization of the process from time $t=0$ to time $t=T>0$. Now the induced probability measures on the space of paths for two distinct values of σ are orthogonal. Therefore likelihood methodology suggests that from a single realization of the process the value of σ can be determined precisely. However, there is a continuity problem here that cannot be easily removed. In practice, it is impossible to make an infinitely precise measurement and so we must allow for the fact that the process actually recorded is a combination of X_t and an error e_t. For example, the combined process, $X_t + e_t$ might be a rounded version of the original Wiener process. However, it can be seen that within any such error neighborhood of X_t lie paths which are not in the support of the Wiener measure. Thus estimates of σ based upon maximum likelihood and the assumption that the error process is negligible will yield inconsistent estimates. A distinction can be made between the inconsistency in this example and that which will be present in a model whose observations are random variables with error. If $X_1, \cdots, X_n$ are i.i.d. normal variables with mean zero and variance σ^2, then the maximum likelihood estimate based upon observations with errors, namely $X_1+e_1, \cdots, X_n+e_n$, will also be inconsistent. However, in this case the size of the inconsistency will be dependent on the size of the errors. In the case of a Wiener process with error, there is no guarantee that the inconsistency in the maximum likelihood estimate will be small when the error process is small. This illustrates the general problem that maximum likelihood estimation for such Wiener processes will not be robust against small deviations from model assumptions.

It might be argued that the problems in the example above appear because of too heavy a reliance on the continuity assumptions that are a part of a Wiener process. To calculate σ on the basis of $X_t, 0 \leq t \leq T$ one typically constructs a partition $0=t_0<t_1<t_2<\cdots<t_{n-1}<t_n=T$ and then uses the fact that $\sum_{i=1}^{n}[X(t_i)-X(t_{i-1})]^2 \to \sigma^2 T$ in the mean as the partition becomes progressively finer. Estimation could be based upon maximum likelihood with a partition that is sufficiently coarse that any errors in the observations have a negligible effect on the inference. However, in practice we will be unable to judge how coarse such a partition need be. It may well be that there is no natural finite set of time points on which to base estimation. (If the process is recorded physically as a graph this will be the case.) In addition, there are cases where the distribution theory associated with a discrete time approximation to a continuous time process becomes very difficult or even intractable. Consider, for example, a gen-

eral birth and death process in continuous time. In order to calculate the marginal distribution of the process associated with observations at discrete time points it is necessary to solve the Kolmogorov equations for the transition probabilities. However, no general solution is available beyond certain special cases such as the pure birth process with linear birthrate.

Thus it can be seen that problems can arise in the use of maximum likelihood both in the continuous time case and in the discrete time approximations. While it is possible that through some judicious selection of observations one can solve these problems on a case by case basis, it is natural to try to develop a theory of inference that works sensibly on the observed process as given and is sufficiently stable as a procedure to be reasonably "indifferent" as to whether a continuous time or discrete time version of the process is used. In situations such as these, the problems of regularity become at least as important as those of efficiency. Our approach here will follow the general philosophy: to find a class of sensible procedures and then to search for an appropriate one within that class. Throughout these pages we have argued for the necessity of imposing regularity on the inference procedure rather than the model. This will become especially important in problems involving stochastic processes where standard regularity conditions (such as constancy of support) cannot be realized on the model. The major hope in such contexts is that inference can be achieved in a space of procedures that is sufficiently restricted as to avoid non-regular methods. One simple restriction that imposes a certain degree of smoothness is that of linearity, and so one might begin by restricting to a linear class of inference functions.

The material in this chapter overlaps with that in a number of recent publications. The best linear unbiased estimators of section 6.1 are found in Grenander (1981), the discrete time examples of section 6.2 are found in a special case in McLeish (1984) and the interesting continuous time examples in Thavaneswaran and Thompson (1986).

6.1. Linear Inference Functions.

We now consider a general process X_t; $t\epsilon[a,b]$. Assume that each X_t is square integrable and that the process is *continuous in mean*, that is that $E|X_s-X_t| \to 0$ as $s \to t$. We begin with the case of an unknown scalar parameter; suppose $E(X_t) = \theta\, m(t)$ where m is a given function and θ is the unknown parameter. Suppose the process X is a (measurable) random element of the space $D[a,b]$ of right continuous functions with left hand limits, this space endowed with the Skorokhod topology (c.f. Billingsley (1968)). We also assume that the function $m(t)$ is a member of this space. Let the dual of $D[a,b]$ be denoted L. This is the space of continuous linear functionals on $D[a,b]$. Now consider a space Ψ of inference functions, closed in the weak-squared topology, of the form

$$\psi(\theta) = L[X_.-\theta\, m(.)] \qquad (6.1)$$

where $L\epsilon\, L$. Suppose that there exists an inference function $\psi^*\epsilon\Psi$ which satisfies the set of *normal equations*

$$E_\theta[\psi^*(\theta)X_s] = m(s) \qquad (6.2)$$

for all $a \leq s \leq b$ and for all θ. We wish to show that ψ^* spans the complete E-sufficient subspace.

Consider an arbitrary $\psi \epsilon \Psi$ and a measure P on $D[a,b]$ such that both $\psi(\theta)$ and $X.$ are integrable. Let $X.^{(i)}, 1 \leq i \leq n$ be independent replications of the process $X.$ with distribution P and let $\hat{P}_n$ be the empirical measure corresponding to these observations. Then by the linearity of the functional L,

$$E_{\hat{P}_n} \psi(\theta) = L[E_{\hat{P}_n} X. - \theta\, m(.)] \tag{6.3}$$

Now by the law of large numbers, since L is linear and continuous, we may take limits in (6.3) as $n \to \infty$ to obtain

$$E_P \psi(\theta) = L[E_P X. - \theta\, m(.)]. \tag{6.4}$$

In particular, replacing P by the measure corresponding to parameter value β,

$$E_\beta \psi(\theta) = (\beta - \theta) L[m(.)]. \tag{6.5}$$

Now suppose ψ^* satisfies (6.2) and $\psi \epsilon \Psi$ is such that $E_\theta[\psi^*(\theta)\psi(\theta)] = 0$. By an argument similar to that above, approximating P_θ by measures with finite support, we can show that

$$E_\theta[\psi^*(\theta)\psi(\theta)] = L\{E_\theta(\psi^*(\theta)[X. - \theta\, m(.)])\} = (1 - \theta)L[m(.)] \tag{6.6}$$

and this is equal to 0 for all θ if and only if $\psi(\theta)$ is E-ancillary. Therefore, the space of all multiples of $\psi^*(\theta)$ is complete E-sufficient.

We now give a specific construction of ψ^* in a special case. Let $R(s,t) = cov_\theta(X_s, X_t)$ be the covariance operator and assume that this is independent of the parameter θ. Denote by ϕ_j, λ_j the orthonormal eigenfunctions, eigenvalues respectively of R, satisfying

$$\lambda_j \phi_j(s) = \int_a^b R(s,t)\phi_j(t)dt, \ j = 1, 2, \ldots$$

Then it follows from the Karhunen-Loève expansion (cf. Grenander (1981)) that

$$R(s,t) = \sum_{j=1}^{\infty} \lambda_j \phi_j(s)\bar{\phi}_j(t)$$

with convergence on the right side occurring absolutely and uniformly.

Denote the coefficients in the expansion of $m(t)$ by $\eta_j = \int_a^b m(t)\bar{\phi}_j(t)dt$. We will

show that the complete E-sufficient subspace is spanned by the function

$\psi^*(\theta) = \int_a^b f^*(t)[X_t - \theta\, m(t)]dt$ where $f^*(t) = \sum_{k=1}^{\infty} \frac{\eta_k}{\lambda_k}\phi_k(t)$ if this series converges uniformly

and absolutely and the resulting function ψ^* lies in Ψ. Note that

$$E_\theta[\psi^*(\theta)X_s] = E_\theta\{X_s \int_a^b f^*(t)[X_t - \theta\, m(t)]dt\} = \int_a^b f^*(t)R(s,t)dt$$

$$= \int_a^b \{\sum_{k=1}^{\infty} \frac{\eta_k}{\lambda_k}\phi_k(t)\}R(s,t)dt = \sum_{k=1}^{\infty} \frac{\eta_k}{\lambda_k}\int_a^b \phi_k(t)R(s,t)dt$$

$$= \sum_{k=1}^{\infty} \eta_k \phi_k(s) = m(s)$$

Therefore the normal equations (6.2) hold.

There is a close relationship between a function spanning the complete E-sufficient subspace and the best linear unbiased estimator of a parameter (c.f. Grenander (1981)). In particular, we have the following simple proposition.

6.1.1. Proposition. Let $T(X)$ be a linear statistic. Then if the function $T(X) - \theta \in \Psi$, $T(X)$ is a best linear unbiased estimator of θ if and only if the set of functions

$\{k(\theta)[T(X)-\theta]\}$ is complete E-sufficient.

Proof. It is well-known that an unbiased linear estimator $T(X)$ is best linear if and only if it is uncorrelated with every linear unbiased estimator of zero, i.e. with every $\psi(\theta)$ such that $\psi\epsilon\Psi$ is E-ancillary.

This proposition allows the computation of a BLUE from the complete E- sufficient generator $\psi^*(\theta)$ whenever it takes the form $k(\theta)[T(X)-\theta]$. Indeed, the BLUE $T(X)$ is obtained as the solution for $\hat{\theta}$ of the equation $\psi^*(\hat{\theta}) = 0$. For example, (c.f. Grenander (1981)) shows in the above context that there exists a unique best linear unbiased estimator $\hat{\theta}$ satisfying the normal equations similar to (6.2) for an estimator.

$$E_\theta[\hat{\theta}X_t] = C(\theta)m(t) \text{ for all } t, \theta$$

6.1.2 Example. Suppose X_t, $0<t<1$ is a Wiener process with drift $\theta\, m(t)$ and variance process $\sigma^2 t$. We wish to estimate the unknown parameter θ. We first seek a solution to the normal equations: consider the general form

$$\psi(\theta) = \int_0^1 f(t)d[X_t - \theta\, m(t)] \tag{6.7}$$

this defined as a stochastic (Wiener) integral. Notice that with $f^*(t) = \dfrac{1}{\sigma^2}\dfrac{d}{dt}m(t)$,

$$E_\theta\{\psi^*(\theta)X_s\} = \int_0^s f^*(t)\sigma^2 dt = m(s).$$

Thus, ψ^* generates the E-sufficient subspace of functions. Setting it equal to 0 and solving, we obtain the best linear unbiased estimator

$$\hat{\theta} = \frac{\int_0^1 m'(s)dX_s}{\int_0^1 [m'(s)]^2 ds}$$

So, for example, when $m(t) = t$, $\hat{\theta} = X(1)$ is the best linear unbiased estimator and

when $m(t) = t^p$, $p > 1$, $\hat{\theta} = \frac{2p-1}{p} \int_0^1 t^{p-1} dX_t = \frac{2p-1}{p} [X(1) - (p-1) \int_0^1 t^{p-2} X_t dt]$.

6.1.3 Example. We include an example of the above technique in the case of a process that is not a martingale. Consider a process on $[0,1]$ with positive definite covariance function

$$R(s,t) = \lambda_0 + 2\sum_{n=1}^{\infty} \lambda_n \cos(2\pi n s)\cos(2\pi n t)$$

where $\lambda_n \geq 0$ for all n and

$$m(t) = 1 + c \, \cos(2\pi t)$$

Then a function spanning the complete E-sufficient subspace is

$$\psi^*(\theta) = \int_0^1 f^*(t)[X_t - \theta \, m(t)]dt$$

where

$$f^*(t) = \lambda_1 + 2\lambda_0 c \cos(2\pi t)$$

and the best linear unbiased estimator of θ is

$$\hat{\theta} = \frac{\int_0^1 f^*(t)X_t dt}{\int_0^1 f^*(t)m(t)dt}$$

6.2 Joint Estimation in Multiparameter Models.

In this section, the parameter θ is assumed a p–dimensional column vector (vectors, unless indicated otherwise, are column vectors). We wish to estimate *jointly* the components of the parameter. The approach to inference here differs somewhat from that in chapter 4 where we also considered multiparameter models. In chapter 4, we considered functions for estimating *components of the parameter vector while regarding the remaining components as nuisance parameters*. In other words, we there discounted the effect of the estimation of components θ_j, $j \neq i$ on the estimation of θ_i.

Consider a space Ψ of inference functions $\psi(\theta;X) = \psi(\theta)$ from the parameter space and the observations into $\mathbf{R}^p$. Note that the equation $\underset{p \times 1}{\psi(\theta)} = \underset{p \times 1}{0}$ establishes p equations in the p unknown components of the parameter vector and may admit a unique solution. As usual, we require that the functions are unbiased, i.e. that $E_P[\psi(\theta(P))] = 0$ and square integrable, i.e.

$$||\psi||_\theta^2 = E_P[\psi^T(\theta)\psi(\theta)] < \infty \tag{6.8}$$

for all $P, \theta = \theta(P)$. The space is also chosen to have constant covariant structure so that the inner product $E_P[\psi^T(\theta(P))\phi(\theta(P))]$ depends on P only through $\theta(P)$. As usual, we also require that Ψ be weak squared closed, i.e. closed in the topology induced by the norms (6.8). The notion of E-ancillarity has a natural extension to the multidimensional case. A function $\phi \epsilon \Psi$ is *E-ancillary* if it is in the weak squared closure of the set of functions ψ such that $E_Q\psi(\theta) = 0$ for all θ, Q. A set S of inference functions is *E-sufficient* if $E_P[\psi(\theta)\phi^T(\theta)] = \underset{p \times p}{0}$ for all $P, \theta = \theta(P)$ implies that ϕ is E-ancillary. Again, the set is complete E-sufficient if this condition is necessary and sufficient.

The local definitions of E-ancillarity and E-sufficiency extend similarly. For example, a function ϕ is k'th order locally E-ancillary if it is in the closure of functions ψ satisfying

$$E_Q[\psi(\theta)] = o[|\theta(Q) - \theta|^k] \quad as \quad \theta(Q) \to \theta.$$

As in the univariate parameter case, the score vector

$$S(\theta) = \begin{bmatrix} \dfrac{\partial}{\partial\theta_1}\log f(x;\theta) \\ \cdot \\ \cdot \\ \cdot \\ \dfrac{\partial}{\partial\theta_p}\log f(x;\theta) \end{bmatrix}$$

is a first order locally E-sufficient function under the standard regularity conditions.

6.2.1 Example. Suppose X_t ; $t=1, 2, \ldots n$ is a real-valued process with p-dimensional parameter $\theta^T = (\theta_1, \theta_2, \ldots \theta_p)$. We assume that the mean function is a linear combination of some known functions with the parameter providing the weight on these functions:

$$E_\theta X_t = \sum_{i=1}^{p} \theta_i m_i(t) \tag{6.9}$$

where m_i are given linearly independent functions and the coefficients θ_i are the unknown parameters. Let R_θ denote the covariance matrix $R_\theta(s,t) = cov_\theta(X_s,X_t)$ and assume that this is a non-singular matrix for each value of θ and depends only on the underlying distribution P through the value of θ.

We consider a space of unbiased, vector-valued inference functions ψ that are linear in the components of X_t, i.e. of the form $\psi(\theta) = \sum_{t=1}^{n} f(t;\theta)[X_t - E_\theta X_t]$ where $f(t;\theta)$ is a function of t and θ into the space of p-dimensional (column) vectors. A function of this form is E-ancillary if $E_\eta \psi(\theta) = 0$ for all θ, η or equivalently if it satisfies

$$\sum_{t=1}^{n} f(t;\theta)m_i(t) = \underset{p\times 1}{0} \quad \text{for all } i,\theta .$$ Then it is easy to see that the subspace of inference functions spanned by $f(t;\theta)$ of the form

$$f^T(t;\theta) = t'th \ row \ of \ R_\theta^{-1}M \tag{6.10}$$

where

$$M = \begin{bmatrix} m_1(1) & . & . & m_p(1) \\ m_1(2) & . & . & m_p(2) \\ . & . & . & . \\ . & . & . & . \\ m_1(n) & . & . & m_p(n) \end{bmatrix} \tag{6.11}$$

is a complete E-sufficient subspace of inference functions. Notice that the estimator obtained by setting these functions all equal to 0 and solving for θ is, in the case that the covariance function $R = R_\theta$ does not depend on θ, the best linear unbiased estimator (see Grenander (1981))

$$\theta^* = (M^T R^{-1} M)^{-1} M^T R^{-1} X \tag{6.12}$$

where $X^T = (X_1, X_2, \ldots X_n)$. In the more general case that the covariance matrix depends on the parameter, we may solve an equation of the form (6.12) by repeated substitution for the root of the E-sufficient inference function.

6.3. Martingale Inference Functions.

In this section, we consider models in which the (conditional) moment structure of a process is specified up to the second conditional moments and we construct a class of inference functions that depend only on this structure. Within this class we select an E-sufficient or locally E-sufficient subspace. We have already seen univariate parameter versions of the first two examples in chapter 5.

6.3.1 Example Suppose $X_i; i=1, 2, \ldots, n$ are square integrable random vectors and X_i is adapted to an increasing family of sigma-fields F_i . Suppose for some p by k matrix of F_{i-1} variates μ_i ,

$$E[X_i | F_{i-1}] = \mu_i^T \theta \qquad (6.13)$$

We consider the space of functions of the form

$$\psi(\theta) = \sum_{i=1}^{n} A_i (X_i - \mu_i^T \theta) \qquad (6.14)$$

where A_i is a matrix of F_{i-1} measurable variates such that the inference function is square integrable. Recall the definitions of E-ancillarity and E-sufficiency for vector parameters described in section 6.2. Suppose the conditional variance

$$var[X_i | F_{i-1}] = \Sigma_i \qquad (6.15)$$

does not depend on the underlying distribution or the value of the parameter and is, with probability one, an invertible matrix. Then if the function

$$\psi^*(\theta) = \sum_{i=1}^{n} \mu_i \Sigma_i^{-1} (X_i - \mu_i^T \theta) \qquad (6.16)$$

is in the space, then the complete E-sufficient subspace consists of multiples $k(\theta)$ of this function.

It should be noted that use of the complete E-sufficient inference function (6.16) assumes that the mean function μ_i and conditional variance function Σ_i are observable. We now obtain a locally E-sufficient function in a slightly more general setting.

6.3.2 Example. The model is the same as that of example 6.3.1 except that in the definition of the space we permit matrices $A_i = A_i(\theta)$ and $\Sigma_i = \Sigma_i(\theta)$ to depend on the parameter of interest (but not on any other nuisance parameters). Then, once again assuming the function is well defined and in the space, the function

$$\psi^*(\theta) = \sum_{i=1}^{n} \mu_i \Sigma_i^{-1}(\theta)(X_i - \mu_i^T \theta) \tag{6.17}$$

spans the complete locally E-sufficient subspace. Moreover, this function is a martingale, and martingale methods therefore apply to obtaining consistency and asymptotic normality of the estimators defined by these functions.

The inference function (6.17) is well known in the literature, defining the *quasi-likelihood* estimators of Wedderburn (1974). There are many possible derivations of this inference function. For example, if we *assumed* that the observations were jointly normal, the function (6.17) is one part of the score function, the part most sensitive to changes in the parameter as reflected in the conditional mean of the observations, and chosen to be unbiased.

6.3.3 Example. Consider stochastic process satisfying a stochastic integral equation of the form

$$X_t = \lambda(\theta)\int_0^t a_s d<M>_s + M_t \qquad (6.18)$$

for a non-random function $\lambda(\theta)$, such that $\lambda(\eta) \neq \lambda(\theta)$ for $\theta \neq \eta$, a predictable process a_t which does not depend on the parameter θ, and a square integrable martingale M_t. We assume that the predictable process a_t is observable and denote by $<M>_t$ the *predictable variation process* which we also assume observable and not dependent on θ. For each θ, denote by $M_t(\theta)$ the process $X_t - \lambda(\theta)\int_0^t a_s d<M>_s$ as θ ranges through the parameter space. Consider the space of all *linear* inference functions of the form

$$\psi(\theta) = \int_0^T H_t dM_t(\theta) \qquad (6.19)$$

Here, H_t is a predictable process independent of the parameter such that the resulting inference function is a square integrable martingale. The stochastic integral is seen to be well defined because $M_t(\theta)$ is a semimartingale for each value of θ. It is not hard to show that the space of such square integrable inference functions is closed in the weak-square topology. It can be seen that when $E_\theta\{\int_0^T H_t a_t d<M>_t\} = 0$, then $\psi(\theta)$ is E-ancillary. So within the space of all functions of the form (6.19), the inference function

$$\psi^*(\theta) = \int_0^T a_t dM_t(\theta) \qquad (6.20)$$

generates the complete E-sufficient subspace.

Remark. Models of the form (6.18) are often written in a more compact differential form $dX_t = \lambda(\theta)a_t d<M>_t + dM_t$. The "differentials" appearing in this *stochastic differential equation* themselves have no meaning. Equations of this sort will simply be shortened forms of the more meaningful integral equation analog.

6.3.4 Example. (diffusion process) Again we denote by W_t the standard Brownian motion process and consider a model of the form

$$dX_t = \lambda(\theta)X_t dt + \sigma dW_t \qquad (6.21)$$

where the *diffusion* coefficient σ is a known, non-random constant and where θ is the unknown parameter that we wish to estimate. The existence and uniqueness of the distribution of a process X_t satisfying an equation of this form subject to an initial condition on the value of X_0 is well known (c.f. Basawa and Rao) and establishes that this is a one-parameter family. As before, we define the space of inference functions of the form $\psi(\theta) = \int_0^T H_t dM_t(\theta)$ where $M_t(\theta) = X_t - \lambda(\theta)\int_0^T X_t dt$. Then within this space of inference functions, the function

$$\psi^*(\theta) = \frac{1}{\sigma^2}[\int_0^T X_t dX_t - \lambda(\theta)\int_0^T X_t^2 dt] \qquad (6.22)$$

generates the complete E-sufficient subspace. The estimator of the parameter θ obtained by solving the equation $\psi^*(\theta) = 0$ is given by

$$\lambda(\hat{\theta}) = \frac{\int_0^T X_t dX_t}{\int_0^T X_t^2 dt} . \qquad (6.23)$$

It turns out that in this example, the above estimator is also the maximum likelihood estimator of the parameter. In order to show this, we must show that the distributions of the processes X_t are all dominated by a single measure so that their likelihoods are well-defined. Then we show that the score function obtained by differentiating the log-likelihood is, in fact, the function (6.22). Alternatively, we might draw on the remark on page 31 but this still requires verification that the score function lies in our restricted space of inference functions. A discussion of maximum likelihood is given by Le Breton (1975). Now the estimator (6.23) is of no practical value unless we are able to approximate it on the basis of observations on the process at discretely many time points $t_i;\ i = 1, 2, \ldots, t_n = T$. The obvious choice is to replace the numerator of (6.23) by its approximant in probability $\sum_i X_{t_i}(X_{t_{i+1}} - X_{t_i})$.

6.3.5. Example. Consider a stochastic process X_t which satisfies a stochastic differential equation of the form

$$dX_t = a_t(\theta)d{<}M{>}_{t\theta} + dM_{t\theta}$$

where, M_t is a square integrable martingale with predictable variation process ${<}M{>}_{t\theta}$ observable up to the value of the parameter θ . Assume also that the predictable process $a_t(\theta)$ is observable up to the parameter value and that there exists a real-valued predictable process $f(t;\eta,\theta)$ such that

$$\int_0^s a_t(\eta)d{<}M{>}_{t\eta} = \int_0^s f(t;\eta,\theta)d{<}M{>}_{t\theta}$$

for all $0{<}s{<}T$ and all η sufficiently close to θ . Notice that $f(t;\theta,\theta) = a_t(\theta)$. As before, define $M_t(\theta) = X_t - \int_0^T a_t(\theta)d{<}M{>}_{t\theta}$ and restrict to a space of inference functions of the form $\psi(\theta) = \int_0^T H_t(\theta)dM_t(\theta)$. We require sufficient regularity to permit interchanging integral and derivative as in the proof below. Then the inference function

$$\psi^*(\theta) = \int_0^T \left[\frac{\partial}{\partial \eta} f(t;\eta;\theta)|_{\eta=\theta} \right] dM_t(\theta)$$

is locally E-sufficient.

Proof. First we observe that for any choice of function ψ and values η, θ,

$$E_\eta \psi(\theta) = E_\eta \{ \int H_t(\theta)[dM_t(\eta)+a_t(\eta)d<M>_{t\eta}-a_t(\theta)d<M>_{t\theta}] = E_\eta \{ \int_0^T H_t(\theta)[f(t;\eta,\theta)-f(t;\theta,\theta)]d<M>_{t\theta} \}.$$

Dividing by $\eta-\theta$ and taking limits as $\eta \to \theta$, we obtain

$$\frac{\partial}{\partial \eta}E_\eta \psi(\theta)|_{\eta=\theta} = E_\theta \int_0^T H_t(\theta)[\frac{\partial}{\partial \theta}f(t;\theta,\theta)]d<M>_{t\theta}.$$

Therefore, the function ψ is locally E-ancillary if this is equal to 0. With ψ^* defined as above, this follows if $E_\theta[\psi(\theta)\psi^*(\theta)] = 0$.

6.3.6. Example (diffusion process) Consider the model

$$dX_t = a_t(\theta)dt + \sigma_t(\theta)dW_t$$

where $a_t(\theta), \sigma_t(\theta)$ are predictable processes, both observable up to the value of the parameter θ. Then in a space of inference functions of the form $\int_0^T H_t(\theta)[dX_t - a_t(\theta)dt]$, the function

$$\psi^*(\theta) = \int\limits_0^T \frac{\partial a_t(\theta)}{\partial \theta} \frac{1}{\sigma_t^2(\theta)} [dX_t - a_t(\theta)dt]$$

is locally E-sufficient.

Note the similarity of the above function to the quasi-likelihood estimating function.

6.4 Applications in Spatial Statistics.

In applications of stochastic modeling to forestry, biology and geography, a spatial scattering of individuals is often modeled as a homogeneous Poisson process that is observed through a "window" which is usually the area under investigation. It is often impossible to observe and record the coordinates of every individual in the process and so the researcher must rely upon incomplete information given by a sampling of the point process. Quadrat sampling is an example of this. Various subregions chosen either randomly and systematically are selected and the total count of individuals within each subregion is recorded. On the basis of these counts, various inferences can be made about the spatial point process including, for example, an estimate of its intensity.

Suppose n such regions are labeled $A_1, \cdots, A_n$ with corresponding point counts $X_1, \cdots, X_n$. For a Poisson process of intensity λ the variables X_i will be Poisson distributed with parameter $\lambda |A_i|$, where $|A_i|$ represents the area of A_i. If the quadrats are selected randomly and the probability of overlap between any two quadrats is small, then $X_1, \cdots, X_n$ are approximately pairwise independent. This suggests that in order to make inferences about λ we could assume that the quadrat counts are mutually independent and apply standard statistical tools. This yields for example the estimator $\hat{\lambda} = (\sum X_i)/(\sum |A_i|)$. However, the distinction between pairwise independence and mutual independence above is not a trivial one. In what follows we apply the theory of inference functions to study the effect of dependence of quadrat counts on inferences. It should be noted that such an analysis has a bearing on other methods of spatial statistics. Typically local descriptions of point processes have the property that the dependence between any two locations vanishes as the locations become more widely separated. Nearest neighbor distributions are an example of this. The Skellam-Moore statistic (see for example Ripley (1981, p. 136)) can be shown to have approximately a chi-square distribution through such approximations by independence. However, in this case, the effect of dependence is non-trivial. Despite the importance of dependence considerations, many spatial statistics are motivated by approximations using product models.

Suppose in the quadrat sampling problem we consider the space of inference functions for λ of the form

$$\psi(\lambda) = \sum_{i=1}^{n} a_i(\lambda)(X_i - \lambda |A_i|) \tag{6.24}$$

where $a_i(\lambda)$, $i=1,...n$ can vary in λ but not in the data. The space $\mathcal{A}$ of E-ancillary functions is the subspace satisfying

$$\sum_{i=1}^{n} a_i(\lambda) |A_i| = 0. \tag{6.25}$$

An elementary argument now establishes that the complete E-sufficient subspace $\mathcal{S}$ is generated by a single function whose coefficients a_i do not depend on λ and satisfy the system of simultaneous linear equations

$$\sum_{i=1}^{n} a_i \, |A_i \cap A_j| = |A_j| \qquad\qquad (6.26)$$

for $j=1,...,n$. The root of this equation is a best linear unbiased estimator for λ. This inference function gives some insight into the distinction between approximate pairwise independence and mutual independence. If $A_1, \ldots, A_n$ are disjoint, then $X_1, \ldots, X_n$ are mutually independent, and $a_1 = ... = a_n$ forms a solution to the equations. For convenience we could set all a_i equal to unity. However even in circumstances where the probability of overlap between randomly chosen A_i and A_j is small the effect of overlap is seen to be sufficiently strong to prevent $a_1 = \cdots = a_n = 1$ from being an approximation to the solution of the linear equations. In this respect, the effect of neighboring quadrats does not vanish as more quadrats are chosen over a larger region of investigation.

An alternative formulation that leads to similar conclusions is to suppose that the homogeneous point process is sampled by nearest neighbor methods at various locations. At each of locations $r_1, \ldots, r_n$ the squared distances from these points to the closest point of the process is observed. Let these squared distances be $u_1,...,u_n$. These will be exponentially distributed with parameter $(\pi\lambda)^{-1}$. However, once again we note that unless the sampling locations are far apart, these variables will not be independent. If we consider a space of inference functions of the form $\sum_{i=1}^{n} a_i(\lambda)[u_i - (\pi\lambda)^{-1}]$ then the complete E-sufficient subspace is once again generated by the function with coefficients a_i that depend on λ and satisfy

$$\sum_{i=1}^{n} a_i \, Cov(u_i, u_j) = 1 \qquad\qquad (6.27)$$

for $j=1,...,n$. The covariances will be functions of λh_{ij} where h_{ij} is the squared distance from r_i to r_j. They can be evaluated numerically as a function of this quantity. The use of functions that are linear in the squared distances rather than the distances themselves is motivated by the fact that if a single nearest neighbor distance is observed (so that there is no problem of dependence between observations) the technique reduces to maximum likelihood estimation. It should be noted however that such estimates will not be robust. A solution to this problem can be found estimating the parameter based upon the quantiles of the distribution. Consider inference functions of the form

$$\sum_{i=1}^{n} a_i \, sgn \left[u_i - \frac{log2}{\pi\lambda} \right] \qquad (6.28)$$

If all the coefficients are chosen to be equal, then the resulting estimator for λ is that value which equates the median of the distribution of the nearest neighbor distances with the empirical median. In Bradley and Small (1986), satisfactory results were obtained by this robust estimate. However, we can hope to do better here by utilizing the covariance structure of the problem. The complete E-sufficient subspace is generated for these functions by that function with coefficients satisfying

$$\sum_{i=1}^{n} a_i \gamma_{ij} = 1 \qquad (6.29)$$

where

$$\gamma_{ij} = \exp\left[-2 \log 2 + \frac{2 \log 2}{\pi} \cos^{-1}[(\frac{h_{ij}\pi\lambda}{4 \log 2})^{1/2}] - \frac{\lambda}{2}[h_{ij}(\frac{4 \log 2}{\pi\lambda} - h_{ij})]^{1/2}\right] \qquad (6.30)$$

when $h_{ij} \leq \frac{4 \log 2}{\pi\lambda}$. If $h_{ij} > \frac{4 \log 2}{\pi\lambda}$ then $\gamma_{ij}=0$. There is some hope that this function achieves a balance between robustness and efficiency for this problem.

Insight into the kinds of methods appropriate for spatial analysis can be obtained by considering inference procedures in more general settings. A spatial process of geometrical objects can be thought of intuitively as a random scattering of particles in Euclidean space. These particles can all be congruent, such as for a *point* process, a process of *lines* or a process of *disks of equal radius*. However, many interesting applications arise in which the particles are not congruent. For example, in metallurgical applications, the flaws in metals can be represented as random scatterings of particles uniformly throughout the metal. The shapes of these particles can be quite irregular. In general the particles may or may not be allowed to overlap, depending upon the particular distributional assumptions that are desired. A realization of such a process can be represented mathematically by a function from a class of subsets of Euclidean space into the non-negative integers. This function, say $n(A)$, simply counts the number of particles of the process which have non-empty intersection with the given set A. If this function is known on a sufficiently rich class of sets A then it is possible to reconstruct the realized process. Again, in practice, only fragmentary information might be available. To distinguish the various sets that appear in the discussion we shall continue to call the sets of the spatial process "particles" and shall now call the sets A on which the count function is recorded "traps". Let C be the class of all traps to be used in the analysis. In the case of quadrat sampling given above, the traps are simply the quadrats themselves, the particles are the points of the Poisson process and the count function the record of the number of points in each quadrat.

An important special case is a Poisson process of lines in the plane. In this case, let C be the set of all convex compact subsets of the plane. Then such sets will have a rectifiable boundary which we denote by ∂C for each compact convex C. Let $|\partial C|$ denote the length of this boundary. Note that a line segment has a boundary whose length is twice that of the line segment as the length of the segment is counted twice (once for each "side"). We construct a homogeneous Poisson process of lines as follows. Let $\mathcal{G}$ be the group of Euclidean motions of the plane, and let M be the set of all lines of infinite extent in the plane. Then $\mathcal{G}$ acts on M through its action on the plane. Up to a constant multiple, there exists a unique measure on M which is invariant with respect to the action of $\mathcal{G}$ and which assigns strictly positive and finite measure to the set of all lines with non-empty intersection with traps C such that $|\partial C| > 0$. For a given trap C_0 we define a probability measure on the set of lines intersecting C_0 by normalizing the measure. If m such random lines are independently chosen to meet C_0 then we call such a process of lines a binomial process. Let C_0 expand to encompass the entire plane while m goes to infinity such that

$$\frac{m}{|\partial C_0|} \to \lambda \tag{6.31}$$

The resulting process of points is called a Poisson process of intensity λ. The number of lines meeting a given trap C will have a Poisson distribution with mean $\lambda |\partial C|$.

In practice, it will be a binomial process that is observed rather than a Poisson process. Returning now to the general setting of a spatial process of geometrical objects (particles), we can construct a space Ψ of inference functions in the following manner. In view of the argument above it seems natural to include in Ψ all functions of the form

$$\sum_{i=1}^{k} a_i(\theta))[n(C_i) - E_\theta(n(C_i))] \tag{6.32}$$

for any set of traps $C_1, \ldots, C_k$ for which the number of particles meeting C_i has finite second moment. Within the class of functions defined as the weak square closure of those above, analysis of the complete E-sufficient subspace can proceed in an analogous manner to that outlined for quadrat counts in a Poisson process, which now becomes a special case.

6.5 Background Details.

Inference for stochastic processes is the subject of a considerable literature and so we confine ourselves to citing that most relevant to the approach of this chapter. The linear approach of section 6.1, 6.2 is inspired by linear regression, the Gauss-Markov Theorem and the best linear unbiased estimators for stochastic processes in Grenander (1981). This book is one of very few sources in which stochastic processes are treated without necessarily using likelihood of pseudo-likelihood functions. The likelihood approach to estimation for stochastic processes can be found in Basawa and Rao (1980) and Hall and Heyde (1980) with the latter concentrating more on the asymptotic behavior of certain estimation procedures.

Section 6.2 is the natural extension of the concepts of chapter 2 to joint estimation in the multiparametric setting. Similar extensions of Godambe's optimality criterion have been developed e.g. Ferreira (1982b). The discrete time examples of section 6.3 are similar to the use of quasi- likelihoods (cf. Wedderburn (1974)) or, equivalently, generalized weighted conditional least squares. They apply, for example, to models involving aggregation (cf. McLeish (1984), Kalbfleisch and Lawless (1984)). Estimating equations obtained from quasi- likelihoods are discussed in Thavaneswaran and Thompson (1986) and Godambe and Heyde (1987) where continuous time models similar to those of section 6.3 are investigated from the point of view of optimality. Classical methods for spatial statistics are investigated in Ripley (1981) and this provides a background for section 6.4.

REFERENCES

Barnett, V.D. (1966) Evaluation of the maximum likelihood estimator where the likelihood equation has multiple roots. *Biometrika* 53, 151-166.

Bartlett, M.S. (1982) The "ideal" estimating equation. *J. Appl. Prob.* 19(A), 187-200.

Basawa, I.V., Prakasa Rao, B.L.S. (1980). *Statistical Inference for Stochastic Processes*. Academic Press, New York.

Basu, D. (1955) A note on the theory of unbiased estimation. *Ann. Math. Statist.* 26, 144-145.

Basu, D. (1958) On statistics independent of sufficient statistics. *Sankhya* 20, 223-226.

Bhapkar, V.P. (1972) On a measure of efficiency of an estimating equation. *Sankhya A* 34, 467-472.

Billingsley, P. (1968). *Convergence of Probability Measures*. Wiley, New York.

Bondesson, L. (1975) Uniformly minimum variance estimation in location parameter families. *Ann. Statist.* 3, 637-660.

Bradley, R. and Small, C. (1986) Statistical analysis of structures at two settlements from Bronze Age England. *MASCA Journal* 4, 86-95.

Chandrasekar, B. and Kale, B.K. (1984) Unbiased statistical estimation functions in presence of nuisance parameters. *J. Statist. Plan. Inf.* 9, 45-54.

Chernoff, H. (1951) A property of some type A regions. *Ann. Math. Statist.* 22, 472-474.

Cox, D.R. and Hinkley, D.V. (1974) *Theoretical Statistics.* Chapman and Hall, London.

Cox, D.R. (1980) Local ancillarity. *Biometrika* 67, 279-286.

Durbin, J. (1960) Estimation of parameters in time-series regression models. *J. Roy. Statist. Soc. B* 22, 139-153.

Efron, B. and Hinkley, D. (1978) Assessing the accuracy of the maximum likelihood estimator: observed versus expected Fisher information. *Biometrika* 65, 457-482.

Efron, B. (1982) Maximum likelihood and decision theory. *Ann. Statist.* 10, 340-356.

Ferreira, P.E. (1981) Extending Fisher's measure of information. *Biometrika* 68, 695-698.

Ferreira, P.E. (1982a) Sequential estimation through estimating equations in the nuisance parameter case. *Ann. Statist.* 10, 167-173.

Ferreira, P.E. (1982b) Multiparametric estimating equations. *Ann. Inst. Statist. Math.* 34A, 423-431.

Ferreira, P.E. (1982c) Estimating equations in the presence of prior knowledge. *Biometrika* 69, 667-669.

Firth, D. (1987) On the efficiency of quasi-likelihood estimation. *Biometrika,* to appear.

Fisher, R.A. (1922) On the mathematical foundations of theoretical statistics. *Philos. Trans. Roy. Soc. London Ser. A* 222, 309-368.

Fisher, R.A. (1925) Theory of statistical estimation. *Proc. Camb. Philos. Soc.* 22, 700-725.

Godambe, V.P. (1960) An optimum property of regular maximum likelihood estimation. *Ann. Math. Statist.* 31, 1208-1211.

Godambe, V.P. and Thompson, M.E. (1974) Estimating equations in the presence of a nuisance parameter. *Ann. Statist.* 2, 568-571.

Godambe, V.P. (1976) Conditional likelihood and unconditional optimum estimating equations. *Biometrika* 63, 277-284.

Godambe, V.P. and Thompson, M.E. (1976) Some aspects of the theory of estimating equations. *J. Statist. Plan. Inf.* 2, 95-104.

Godambe, V.P. (1980) On sufficiency and ancillarity in the presence of a nuisance parameter. *Biometrika* 67, 269-276.

Godambe, V.P. (1984) On ancillarity and Fisher information in the presence of a nuisance parameter. *Biometrika* 71, 626-629.

Godambe, V.P. and Thompson, M.E. (1984) Robust estimation through estimating equations. *Biometrika* 71, 115-125.

Godambe, V.P. (1985) The foundation of finite sample estimation in stochastic processes. *Biometrika* 72, 419-428.

Godambe, V.P. and Thompson, M.E. (1986) Parameters of superpopulation and survey population: their relationships and estimation. *Int. Statist. Rev.* 54, 127-138.

Godambe, V.P. and Heyde, C.C. (1987) Quasi-likelihood and optimum estimation. *Int. Statist. Rev.* 55, to appear.

Grenander, U. (1981). *Abstract Inference.* Wiley, New York.

Hall, P. and Heyde, C.C. (1980) *Martingale Limit Theory and its Application.* Academic Press, New York.

Kalbfleisch, J.D. and Lawless, J.F. (1984) Least squares estimation of transition probabilities from aggregate data. *Canad. J. Statist.* 12, 169-182.

Kale, B.K. (1961) On the solution of the likelihood equation by iteration processes. *Biometrika* 48, 452-456.

Kale, B.K. (1962a) An extension of Cramer-Rao inequality for statistical estimation functions. *Skand. Aktuar.* 45, 80-89.

Kale, B.K. (1962b) On the solution of likelihood equations by iteration processes: the multiparametric case. *Biometrika* 49, 479-486.

Kendall, M.G. (1951) Regression, structure and functional relationship I, *Biometrika* 38, 11-25.

Kimball, B.F. (1946) Sufficient statistical estimation functions for the parameters of the distribution of maximum values. *Ann. Math. Statist.* 17, 299-309.

Klimko, L.A. and Nelson, P.I. (1978) On conditional least squares estimation for stochastic processes. *Ann. Statist.* 6, 629-642.

Le Breton, A. (1975) Estimation de paramètres dans un modèle de système dynamique à état et observation régis par des équations différentielles stochastiques. *C. R. Acad. Sci. Paris Sér. A-B* 280, 1377-1380.

Lehmann, E.L. and Scheffe, H. (1950) Completeness, similar regions and unbiased estimation. *Sankhya* 10, 305-340.

Lehmann, E.L. and Scheffe, H. (1955) Completeness, similar regions and unbiased estimation. *Sankhya* 15, 219-236.

Lehmann, E.L. and Scheffe, H. (1956) Correction. *Sankhya* 17, 250.

Lehmann, E.L. (1981) An interpretation of completeness and Basu's theorem. *J. Amer. Statist. Assoc.* 76, 335-340.

Lindsay, B. (1982) Conditional score function; some optimality results. *Biometrika* 69, 503-512.

McCullagh, P. (1984) Local sufficiency. *Biometrika* 71, 233-244.

McLeish, D.L. (1983) Martingales and estimating equations for censored and aggregate data. *Technical Report Series of the Laboratory for Research in Statistics and*

Probability #12, Carleton University.

McLeish, D.L. (1984) Estimation for aggregate models: the aggregate Markov chain. *Canad. J. Statist.* 12, 256-282.

Morton, R. (1981) Efficiency of estimating equations and the use of pivots. *Biometrika* 68, 227-233.

Morton, R. (1981) Optimal estimating equations with applications to insect development times. *Aust. J. Statist.* 23(2), 204-213.

Okuma, A. (1975) Optimal estimating equations for a model with nuisance parameter. *Tamkang J. Math.* 6, 239-249.

Okuma, A. (1976) On invariance of estimating equations. *Bull. Kyushu Inst. Tech. (M and Ns)* 23, 11-16.

Okuma, A. (1977) Some applications of partly sufficient statistics to estimating equations in the presence of a nuisance parameter. *Bull. Kyushu Inst. Tech. (M and Ns)* 24, 29-36.

Pearson, K. (1894) Contributions to the mathematical theory of evolution. *Phil. Trans. Roy. Soc. Ser. A* 185, 71-110.

Pfanzagl, J. (1970) On the asymptotic efficiency of median unbiased estimates. *Ann. Math. Statist.* 41, 1500-1509.

Rao, C.R. (1952) Some theorems on minimum variance estimation. *Sankhya* 12, 27-42.

Reeds J.A. (1985) Asymptotic number of roots of Cauchy location likelihood equations. *Ann. Statist.* 13, 775-784.

Ripley, B.D. (1981) *Spatial Statistics.* John Wiley and Sons, New York.

Skovgaard I.M. (1985) A second order investigation of asymptotic ancillarity. *Ann. Statist.* 13, 534-551.

Small, C.G. and McLeish, D.L. (1988) Generalizations of ancillarity, completeness and sufficiency in an inference function space. *Ann. Statist.* 16, to appear.

Sprott, D.A. and Viveros-Aguilera, R. (1984) The interpretation of maximum likelihood estimation. *Canad. J. Statist.* 12, 27-38.

Thavaneswaran, A. and Thompson, M.E. (1986) Optimal estimation for semimartingales. *J. Appl. Probab.* 23, 409-417.

Wedderburn, R.W.M. (1974). Quasi-likelihood functions, generalized linear models, and the Gauss-Newton method. *Biometrika* 61, 439-447

Whittle, P. (1961) Gaussian estimation in stationary time-series. *Bull. Int. Statist. Inst.* 39, 1-26.

Wilks, S.S. (1938) Shortest average confidence intervals from large samples. *Ann. Math. Statist.* 9, 166-175.

Wilks, S.S. and Daly, J.F. (1939) An optimum property of confidence regions associated with the likelihood function. *Ann. Math. Statist.* 10, 225-235.

INDEX

Lecture Notes in Statistics

ctd. on inside back cover

Lecture Notes in Statistics